INVENTORS AND DISCOVERERS

Changing Our World

INVENTORS AND DISCOVERERS

Changing Our World

National Geographic Society

INVENTORS AND DISCOVERERS
Changing Our World

Published by
The National Geographic
Society

Gilbert M. Grosvenor
*President and
Chairman of the Board*

Owen R. Anderson
Executive Vice President

Robert L. Breeden
*Senior Vice President,
Publications and
Educational Media*

Prepared by
National Geographic
Book Service

Charles O. Hyman
Director

Ross S. Bennett
Associate Director

Margaret Sedeen
Managing Editor

Susan C. Eckert
Director of Research

Staff for this book

Elizabeth L. Newhouse
Editor

Linda B. Meyerriecks
Greta Arnold
Illustrations Editors

David M. Seager
Art Director

Cathryn P. Buchanan
Chief Researcher

Jennifer Gorham Ackerman
Mary B. Dickinson
Carole Douglis
Carol Bittig Lutyk
David F. Robinson
Jonathan B. Tourtellot
Lynn Addison Yorke
Editor-Writers

Ratri Banerjee
Gretchen C. Bordelon
James B. Enzinna
Catherine Herbert Howell
Melanie Patt-Corner
Jean Kaplan Teichroew
David W. Wooddell
Editorial Researchers

Mary Grady
Geography Intern

Lise Swinson Sajewski
Style

Robert Friedel
Associate Professor of
History
University of Maryland
Chief Consultant

Charlotte Golin
Design Assistant

Sara Grosvenor
Caroline Sheen
David W. Wooddell
Illustrations Researchers

Laurie A. Smith
Illustrations Assistant

Karen F. Edwards
Traffic Manager

R. Gary Colbert
*Senior Administrative
Assistant*

Teresita Cóquia Sison
Editorial Assistant

Dianne L. Hardy
Lisa S. Jenkins
Indexers

John T. Dunn
Technical Director

Richard S. Wain
Production Manager

Andrea Crosman
Production Coordinator

Leslie A. Adams
Emily F. Gwynn
Production Assistants

David V. Evans
Quality Control Manager

Leslie Allen
Leah Bendavid-Val
James A. Cox
Jerry Camarillo Dunn, Jr.
Marguerite Suarez Dunn
Stephen S. Hall
Edward Lanouette
Joyce B. Marshall
Anne Meadows
Shelley L. Sperry
Susan J. Swartz
Anne E. Withers
Contributors

First edition: 225,000 copies
320 pages, 356 illustrations.

Discoverers & Inventors

By Daniel J. Boorstin

Man, we are told, is the inventing animal. He is also the discovering animal. Through most of history these two features of human nature have seemed quite separate. Discovery has been everybody's everyday experience. We live and learn—which means we constantly discover. But inventions have been rare and far between.

World discoverers have advanced inch by inch, gradually across deserts, slowly down the coasts of Africa, island by island across the Pacific. Every age adds its own discoveries to the legacy handed down by its grandfathers. While history has been punctuated by the great discoveries—of the Earth's path around the sun, of passageways around the world, of the constellations within the atom—the processes of discovery have been unceasing and incremental. Seldom has anything once discovered—the shape of a continent, the source of a river, the anatomy of the atom—ever been forgotten or successfully concealed.

But few ages have seen great world-transforming inventions. The incremental inventions that were being made throughout history left few records. Speech and language and writing, the wheel and the sail, the stirrup and the plow, which marked epochs in the rise of civilization, were usually invented anonymously over ill-defined millennia.

When I selected great discoverers in Western civilization for my book *The Discoverers,* I had to make invidious choices. For every little advance along a risky, half-known path, every mountain first climbed, river source found, desert crossed or island mapped, has enlarged human life and opportunity. Yet it was possible to tell the story of great Western waves of discovery with only occasional mention of the world-changing inventions. This reminded me of the remarkable separation of discoverers from inventors during most of history. We all know the names of the great discoverers—from Democritus and Aristotle to Copernicus, Galileo, Columbus, Pasteur, and Einstein. But only the historian of technology can name the great inventors before recent times.

Time—the primordial dimension of history—was charted and measured by the instruments of anonymous inventors. The technology for measuring time—the water clock, the sundial, the pendulum, the mechanical clock, and the sailors' instruments for taking bearings by the sun and the stars—made possible the epochal discoveries of Copernicus and Galileo. But these elementary time-measuring devices were invented and perfected over centuries. More often than not they were communal achievements. People agreed on the definitions of the sundial, recognized the chiming of the clock on the square or the bell in the church tower, and shared the experiences of sailors using cross-staff and astronomical tables to help refine latitude. Those who defined the instruments of timekeeping—Christiaan Huygens, William Clement,

Robert Hooke, and John Harrison among others—do not enter our history books until recent centuries. Meanwhile, whole communities had come to share the accumulating benefits of the improving technologies of time measurement—in the promptness of private and public meetings, in the regularity of church worship, in the shared fruits of adventuring seafarers and scrupulous cartographers. We cannot know whom to thank for these conveniences.

The classic chronicle of discovery is anything but anonymous. On the contrary, it is intensely, even intimately, personal. Columbus's achievement was a work of salesmanship, courage, expertise, knowledge of wind and current, organizational ability, and stick-to-itiveness. He did have a compass and could find his bearings by ancient methods. But his was a personal, not a technological, triumph. Nobody before him had just the qualities, or found quite the opportunity, to do it.

To make sense of the story of the great discoverers, then, I needed to chronicle only a few crucial inventions. These included the clock, the compass, the printing press, the telescope, and the microscope. The contrast, too, between where the great discoverers and the great inventors commonly did their work is quite striking. Both were a varied and unpredictable lot. But we can say that the great discoverers—openers of vistas—more often than not worked in the open air. Of course among discoverers too there were conspicuous exceptions—the Galens and Harveys. Still, generally speaking, discoverers were darers—who rode the oceans, climbed the mountains, peered into the heavens. Even the great works of discovering antiquity were risky. They required digging amidst suspicious inhabitants in the ruins of Mycenae and Troy, in the tomb of Tutankhamun, or the windswept deserts of Mesopotamia.

Inventors, in sharp contrast, were commonly artisans. Their habitat was the stuffy workshop where they forged axles, cut cogs on wheels, polished lenses and mirrors. While more often than not the discoverer had his eyes set on a distant horizon of space or time, the inventor commonly was myopic. He focused sharply, measured closely, so he could fit all the parts together. The inventor shaped what he held in his hand. The weakness of the one was far-sight, overlooking nearby obstacles, of the other was near-sight, ignoring the long vistas.

Until modern times, these were two worlds with hopes and visions of quite different dimensions. The great discoverer, the Columbus, needed the sponsorship of sovereigns to finance expeditions, to defend voyages and finds against unknown savages and unfriendly rivals. The

A Foucault pendulum swings as the Earth turns beneath it. At the National Museum of American History in Washington, D. C., the 240-pound gold-plated hollow bob hangs on a cable and topples pegs in a seemingly clockwise rotation that is an illusion—French physicist Jean Foucault invented this pendulum in 1851 to prove for the first time in a laboratory that Earth rotates counterclockwise.

great inventor was apt to be an artisan working secure within his guild.

An unsung revolution, more basic even than the industrial revolution recounted in this volume, has united these two worlds. In modern times, as never before, adventurer and artisan have been assimilated into a single community of questing mankind. Western progress since the 18th century has dramatically mixed and unified their roles. Modern discoverer and modern inventor have been drawn together by forces beyond their control in an undeclared, sometimes involuntary, partnership. Of course the restless personalities vary as much as ever. But now inventors have been taken out into the open air—to the deep ocean bottom and the oxygen-rare mountaintops. At the same time discoverers have walked into the wilderness of the laboratory.

This new unlimited partnership can be dated from about the period with which this volume opens—from the beginning of the 18th century. "Scientists," a new word, signaled the new partnership. Surprisingly, its first recorded use did not come until 1840. "We need very much a name," said the brilliant English philosopher-mathematician William Whewell (1794-1866), "to describe a cultivator of science in general. I should incline to call him a Scientist." About the same time the realm of the artisan was being transformed, and the word "artist" came into use to describe "one who practises a manual art in which there is much room for display of taste; one who makes his craft a 'fine art.' " When the inventor-craftsman was naturalized into the realms of science, his aesthetic role was taken from him to be added to the artisans in the fine arts. A prophetic writer in *Blackwood's Magazine* in 1840 explained, "Leonardo was mentally a seeker after truth—a scientist; Correggio was an assertor of truth—an artist."

The modern community of the "cultivator of science in general" was even more novel and more revolutionary than Whewell imagined. This new age might have been christened the age of invention, just as its predecessor in the 15th century was the age of discovery. But it would have been misleading, by perpetuating the disappearing distinction.

The new era signaled an unprecedented symbiosis. Of the few earlier world-transforming inventions, one in particular symbolized and declared the new unity of artisans and knowledge seekers. This was, of course, the new art of printing. Gutenberg and his main partner, Fust, who were only goldsmiths, managed to devise "the art preservative of all the arts." ("*Ars artium omnium conservatrix.*") Printing, a humble, simple invention, became the whole world's vehicle of discovery. Every age now had the power to inherit the whole world's legacy of discoveries—and leave its legacy to all later times and places. What better symbol than printing for the modern collaboration of inventors and discoverers? Succeeding centuries would cement this collaboration—which climaxed in the achievements described in this book.

The modern inventors' transformations of experience were usually by-products. These unintended, outreaching, everyone-touching consequences were possible because inventors took possession of three new areas. They assumed at least three new roles. They became devisers of new materials. They became makers of new sources of energy.

A Julian calendar commissioned by the French duc de Berry in the early 1400s shows new moons and the zodiac signs. As the sun god Phoebus beams from his chariot, peasants bask at the hearth in February (above) and mow hay in June (opposite).

For thousands of years, people have counted the days and organized them into weeks, months, and years. Invented by many cultures to keep track of the times for planting and harvesting or holding religious ceremonies, calendars now help coordinate every type of human activity. The Gregorian calendar, used in most of the world today, was first adopted in 1582.

And they crafted new instruments for sharing experience.

The prosaic innovations in mining and metallurgy during the industrial revolution deserve far more dignity than historians usually give them. For millennia alchemists had tried to turn base metals into gold. But the new ways of mining and of making iron more versatile—and then of compounding iron into steel—were worth many gold mines. Iron and steel would make possible the devices that mined the coal to produce more iron and steel. And these became the skeletons for the machines of mass production. Later the myriad new plastics would multiply all these possibilities a thousandfold. The innovators faced the troublesome but exhilarating problem of inventing names—from Bakelite to nylon—for materials that God had never created.

The novelty of the steam engine was less in its new power than in the outlandish idea that man could make and harness new kinds of power. A melodramatic symbol—of the new unity of discovery and invention, of the new outdoor habitat of the inventor, and of the sharing of risks between inventor and discoverer—was Benjamin Franklin with his electrical kite and his lightning rod. In a single dangerous experiment he identified a new force. At once he provided a new device for the protection, propulsion, and illumination of mankind. The bizarre possibility of new kinds of energy became less bizarre with the elaboration of electrical power and the invention of the internal-combustion engine.

Most outlandish of all was the discovery that the power in the "unbreakable" atom could move factories, transform warfare, unbalance the balances of power among nations, and might even shatter the planet. The daily news dramatized the obsolescence of old distinctions between artisan and adventurer. The world had become an open-air laboratory, and the scientist's workshop had become a jungle of discovery. It was not easy to calculate or control the by-products of the new discoverer-inventor partnerships. Would the new forces of energy—mechanical, chemical, nuclear—produce new forms of waste? Where to put them? Or perhaps try to invent new uses for them?

Another gargantuan and curious new power distinguished the new-age inventions. They no longer merely increased man's capacity to make or improve familiar products—to dig, and cut, and shape, and move, and build, and destroy. They created a strange new set of tasks from novel ways to share experience. Photography and electronics captured and diffused the visual image, the heard voice, the sensations of moving and of being encompassed. New devices assimilated places and times, recaptured the moving images of the dead, the invisible, and the remote. The inventor had unwittingly become an impresario, playing the role of public discoverer with his camera, telephone, phonograph, radio, motion-picture camera, projector, television set, or VCR. He had become an opener and amplifier of worlds for the millions.

New technologies were doing for all experience what printing did for knowledge. Steam transportation moved commodities and the people who made them across oceans. Electricity enlarged the day and illuminated the night. Steam power eased the production of electricity, and electronics carried information, ideas, and images everywhere.

The world became a discovering community. The discoverer in a space capsule en route to the moon came into everybody's living room. Timid stay-at-homes shared the dangerous moments. A whole nation could share the suspense, the hope, and the terror of the great ventures, at the moment of triumph—landing on the moon—or of tragedy—the explosion of the *Challenger*. Would the modern Ulysses and Columbuses and Magellans still be heroes? We all were there!

The new partnership of the search to know—discovery—and the passion to innovate—invention—made it newly impossible to prophesy impossibilities. We were losing our capacity for suspense and surprise. Some new light material, some new propellant, and some new airplane design had broken the sound barrier! After the moon there was Mars, and after Mars . . . ? Was there any limit to the capacity of future "generations"—of computers? In recent history had not the most expert predictors of impossibility (like Rutherford on the atom) been proved outrageously wrong? Every invention widened vistas and so multiplied incentives and opportunities for the "cultivator of science in general." Naysayers found their scripts destroyed. Cassandras were obsolete. One after another, impossible discoveries (the sensation of walking weightless on the moon) were accomplished by impossible inventions (computer technology). The word "impossible," which sententious moralists from Epictetus to William Ernest Henley and Rudyard Kipling had tried in

vain to banish from the young man's lexicon, lost its meaning in the new community of inventors and discoverers. Dared man at long last say, "I am the master of my fate"?

Every day invention was serving discovery in new ways. Glass display windows enticed consumers with new products. Special glass for scientific instruments, which survived the highest and the lowest temperatures, sharpened vision in the laboratory. Photography with novel sensitivities captured and diffused images of what was once invisible. Electronic technology and computers multiplied miracles of everyday discovery. Atomic energy and lasers catapulted into the unimagined.

This exhilarating collaboration has brought new temptations without destroying the old. Discoverers continue to be menaced by the illusion of knowledge. Inventors continue to be menaced by the terminal fallacy, the illusion that a final improvement has been made.

The framers of our federal Constitution prophetically glimpsed this partnership when it was just beginning. They gave to Congress (Art. I, Sec. 8) the power by laws of patent and copyright "to promote the Progress of Science [discovery] and useful Arts [invention]." It was this emerging community of the "cultivators of science in general" for whom our 20th century would create new institutions. Universities became fertile centers. "Research and Development" entered our language as businesses risked billions on the new partnership. An odd couple—the courageous heroic Ulysses and the dogged prosaic Edison—were cemented in a partnership unlikely ever to be dissolved.

Early civilizations found ways to measure years and group days into weeks, but it was not until the 14th-century invention of the mechanical clock that people learned to standardize the day. By the century's end, Europeans could probably count minutes and hours in uniform increments. This freed them from dependence on the sun and fostered belief in a world where quantitative measurement and mathematical certainty could be applied to nature—thus heralding the modern age of science and industry.

Since then, the clock has served as a model of precision and accuracy for all machines. Portable timepieces powered by uncoiling springs appeared in the early 16th century. In the 1600s accuracy received a boost when pendulums began regulating clock mechanisms. In the 1800s cheap pocket watches made punctuality possible for most people, and railroads and factories adopted schedules. Today, digital watches and clocks regulated by electricity, quartz crystals, or oscillating atoms dictate the tempo of life in industrialized nations.

The Power of Steam

By Margaret Sedeen

Oliver Evans was a Yankee and an inventor with a flair for promotion. He designed a flour mill with four stories of automated machinery and sold licenses to people who wanted to copy the system. George Washington bought one after the war and built an Evans mill at Mount Vernon. In 1805 Evans constructed a 30-foot-long steam-powered vehicle to dredge garbage and trash from the Philadelphia waterfront. He named his amphibious digger *Orukter Amphibolos,* advertised it in the *Philadelphia Gazette,* and charged curious persons 25 cents each to gape at the scow-on-wheels as it streamed sparks and smoke, shrieking and lumbering through the streets and into the Schuylkill River. It was America's first powered land vehicle.

In 1813 Evans prophesied that before the United States was a generation older, its citizens would be riding in newfangled carriages, leaving Washington, D. C., before breakfast and arriving in New York by suppertime. Birds, said Evans, would scarcely fly faster than passengers on these "stages moved by steam engines."

People said he was crazy. But he was right. His prediction came true only 25 years later.

What else would a steam engine do? Saw timber, pump water, work a rolling mill or a forge hammer, propel a boat, raise coal from a mine, chop grain, cut stone, clean and grind rags for a paper mill, even make ice. All this and more, promised Oliver Evans, would come from "the power of steam."

This ephemeral collaboration of fire, air, and water, these puffs, wisps, and roils of vapor, when confined and intensified by the genius of inventors became a force that drove wheels in revolutions both mechanical and social. Steam did not initiate the cataclysm known as the industrial revolution. Since before the Middle Ages, mines, mills, workshops, and forges had been mechanized by wind and water, and by the muscles of animals and human beings. But in the late 18th century the power of steam energized the world of work, especially in Great Britain, and sped the industrial revolution to its dynamic climax.

The impetus came from the depths of the British coal mines. By 1650 Great Britain led all other nations in mining and manufacturing. A thriving textile industry developed new methods of spinning and weaving that demanded new materials from which to make machines and new sources of power.

Iron and coal met the demand. By 1700 Britain was mining three million tons of coal a year, perhaps six times more than the rest of the world together. A hundred years later, the output was ten million tons, and in 1900 it was seventy-five million. Steam power made this growth possible and it made it necessary since steam power

helped mine coal and consumed enormous quantities as fuel.

The iron industry saw a similar expansion. The two enterprises had become inextricably linked in 1709 when Abraham Darby, the patriarch of a family of Quaker forgemasters in the Midlands of Britain, began producing iron in a blast furnace that smelted with a coke fire instead of the charcoal that iron forges had used for centuries. His son, also Abraham, devised improvements in the smelting process that increased production and gave a purer iron. The third Abraham Darby developed new uses for the products of their Coalbrookdale foundry in the Severn River Valley. One of the earliest uses was the casting of iron cylinders for the atmospheric steam engine invented in 1712 by an Exeter metalworker and ironmonger named Thomas Newcomen.

Early engines like Newcomen's are called "atmospheric" because they depended on a property of the atmosphere recognized in the late 17th century by the French inventor, Denis Papin. Papin realized that air contains matter and thus can exert pressure. He also understood the concept of a vacuum, that is, the absence of matter. And he saw that when air is introduced into a vacuum, the pressure it exerts can do work, can move objects.

In 1690 Papin constructed a model to demonstrate his theory. The little machine had a brass cylinder two and a half inches in diameter and a piston that, in spite of the best metalworking skills of the time, did not fit very well. Papin put into the cylinder about a third of an inch of water, heated it with a fire below the container, and watched the steam raise the piston. A catch held it up while Papin moved the cylinder from the fire. As the cylinder cooled, the steam condensed, creating a vacuum. Then the engineer released the catch to allow the pressure of the atmosphere to lower the piston.

Although it was tiny, when this "engine" was attached to a pulley, it could lift a 60-pound weight. After each stroke of the piston, Papin had to cool the cylinder, put in more water, and start over. The process was slow and awkward, and the machine never went beyond the inventor's single model. But the principle was the basis for the next developments in the power of steam.

Relegated to history, the William Mason *sits on the turntable in the Baltimore & Ohio Railroad Museum's roundhouse, where workers once cleaned and repaired steam engines. The 1856 locomotive belongs to the age of steam, when inventors harnessed the power of water vapor to change the world. In the space of two centuries, steam engines modernized mining, manufacturing, and agriculture and revolutionized transportation. Vehicles from the paddle wheeler to the ocean liner, from the* Best Friend of Charleston *to the* Orient Express, *once hauled passengers and goods to every corner of the planet.*

Britain's great appetite for coal, used for heating as well as for industry, depleted the shallow, easily worked veins that had been mined since the 13th century. By the late 17th century coal mines, and metal mines as well, were going deeper and deeper. As the shafts reached below the water table, tunnels regularly flooded, and owners had to abandon mines that could not be drained by old methods, such as hand pumps or continuous chains carrying buckets or rags.

Engineers and mechanics searched for practical ways to pump water from deep mines. The first solution came from a Devonshire man, one Thomas Savery, known as Captain—whether because he was a military engineer or a sea captain or a Cornish mining engineer (they were called captain then), no one knows. Whatever he was, Captain Savery had friends at Court and was a Fellow of the Royal Society, where he met leading gentleman-scientists of his day. Savery was the owner of many patents. His association with people of influence undoubtedly helped him, since the granting of a patent required an Act of Parliament and the assent of the monarch. Savery's most important patent, in 1698, was for a device that he named the Miner's Friend. It was "a new Invention for Raiseing of Water and occasioning Motion to all sorts of Mill Work by the Impellent Force of Fire, which will be of great Use and Advantage for Drayning Mines, Serveing Towns with Water, and for the Working of all Sort of Mills where they have not the Benefitt of Water nor Constant Windes."

Parliament extended Savery's patent to give him 35 years of protec-

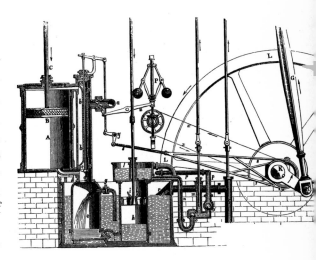

An idealized portrait of young Scotsman James Watt and his attentive family (opposite) perpetuates the legend of the boy's fascination with steaming teakettles. Schooled in his father's shipbuilding workshop, Watt grew up to be a skilled instrument maker. In 1763, at age 28, he was asked to repair a model of a steam engine designed by Thomas Newcomen to pump water from flooded coal mines. The engine was simple: A large piston inside a cylinder was raised and lowered by alternating blasts of steam and cold water. Watt fixed the model, then went on to design an engine that was more efficient in that it cooled the steam in a separate condenser and thus saved energy. He then made the engine more versatile by using interlocking gears to convert the up-and-down motion of the piston to rotating motion.

Side by side, old-fashioned horse power and Newcomen's steam power (top) aided coal miners in fueling the industrial revolution. By 1790 James Watt's double-acting rotative engine (above) turned the wheels of flour mills, textile mills, and iron foundries throughout Great Britain.

tion for the principle of raising water by means of fire and for his machine, which he called a fire engine, though it was only a pump. It had no piston. It fed steam from a boiler through a pipe into a flask-shaped receiving container. This was connected to one pipe that stood in the water to be moved and to another that carried the water out the top. One-way valves helped control the water direction. A flow of water on the outside of the flask cooled it and condensed the steam, creating a vacuum which sucked water up in the first pipe and into the flask. Steam from the boiler drove it through the second pipe.

Savery advertised his engines, and they went into service pumping water for houses of the gentry and, in factories, returning water to waterwheels, but, so far as is known, never in a mine. They could only raise water about 50 feet. This meant that at least one series of five or six pumps, one above another, would have been needed for each working pit in a mine 300 feet deep. Besides, in that era there were no materials or workmanship to make strong boilers. In tests to increase pressure (and thus pumping power), Savery's boilers had a way of violently coming apart at the seams.

It was Thomas Newcomen, a Baptist lay preacher and a plain man without friends in high places, who built the first true atmospheric steam engine in 1712. Newcomen's design was very different from that of Savery's machine, and far superior, but by the broad scope of the 1698 patent, no one in England, Wales, or Scotland but Captain Savery alone had the right to raise water by fire. Working under license from

A cotton weaver in Burnley, England, wears her hair in a bun to avoid snagging it in a belt drive. Steam power kept these belts turning and the looms clacking into the late 1970s. Factories arose here in Lancashire during the 1700s, when weavers abandoned their hearthside handlooms for mechanized equipment at the newly built textile mills. As mass production replaced piecework, cotton cloth—once a luxury—became more widely available. Steamships carried the fabric across the Atlantic Ocean to merchants like L. S. Driggs, whose Lace and Bonnet Store (above) sold elegant dry goods to the ladies of Boston in the 1850s.

Savery, Newcomen produced engines that were praised as "the beautifullest and most perfect" ever seen. After Savery's death in 1715, his heirs, to their own profit, manufactured and leased Newcomen engines. Even on the continent of Europe, the Newcomen engine was widely used and admired, but Thomas Newcomen himself lived and died known only to a few associates and profited little from an invention that was essential to the culmination of the industrial revolution.

Like most inventors, Newcomen combined elements of existing technology in an ingenious way and then added something of his own. His machine had a boiler, a cylinder, and a piston. The top of the piston was hooked to a wooden beam at the opposite end of which was a connection to a pump rod heavy enough to hold the piston raised. Steam was injected from the boiler to the cylinder at the pressure of the atmosphere. A jet of cold water into the cylinder condensed the steam, the piston dropped, and the beam seesawed to lift the pump rod. Another burst of steam beneath the piston raised it to begin again.

A typical Newcomen engine was an impressive sight and a prodigy of noise. The cylinder, poised above a domed boiler five and a half feet in diameter, was almost eight feet high. The beam, centrally balanced on a sturdy wooden frame or on a brickwork structure like the wall of a house, reared above the heads of workmen on the ground. Eight or ten times a minute, the descending pump rod sighed loudly, bumped as it hit bottom, and in the raising, creaked, wheezed, sucked, and bumped again as it came to rest at the top of its stroke, while ten gallons of water gushed from the mouth of the pipe. There was even a valve named for the noise it made: snift, snift, snift.

The motion of the beam, rather than a human operator, opened and shut the valves that emitted jets of steam and water. This system was one of Thomas Newcomen's particular contributions to the technology of steam power because it made the action automatic. Newcomen's engine pumped from far greater depths than Savery's, and it didn't blow up because it could do its work with steam equal to the pressure of the atmosphere. While the injection of cold water into the cylinder was a clever idea, the need to heat and cool the cylinder with every stroke made a machine of abysmally low efficiency.

Engineers who followed made improvements. The carefully machined cylinders and other refinements of John Smeaton brought the Newcomen engine to its peak of performance, but even so it consumed huge amounts of fuel in proportion to the amount of work it did—not so important in a mine full of coal, but important enough if the steam engine was ever to find widespread use outside the collieries and a few metal mines where power was worth whatever it cost.

Almost a century after Newcomen's death in 1729, a group of Manchester workmen assembled one evening to hear a lecture on the application of science to the work of skilled artisans. It was the view, in those days, of certain British statesmen, mine and factory owners, and financiers that the working classes—or at least the clean and respectable among them—would benefit from a little education. The gentlemen felt that carpenters or machinists or dyers would be eager to put in their

In 1839 James Nasmyth, a builder of machine tools and locomotives in Manchester, England, designed a vertical steam hammer—and then painted its portrait (opposite). Powerful enough to forge iron beams and gentle enough to crack eggshells, the hammer earned a place of honor at London's 1851 Crystal Palace Exhibition.

For the 1876 U. S. Centennial Exhibition in Philadelphia, American inventor George Corliss constructed a 40-foot-high steam engine (above) to supply power for all 8,000 machines in Machinery Hall. Improved regulators in Corliss's engine made it much more efficient than earlier models.

12 hours on the job and then spend a few more listening to edifying speeches on the chemistry of bleaching textiles or the Carboniferous formation of peat. They weren't, but before their disinterest became obvious, these schools had spread all over England and Scotland, and there were several in the United States.

On this occasion, the opening of the Manchester Mechanics' Institution in 1825, a banker named Benjamin Heywood cited the single phenomenon of the age most apt to his point: "There cannot indeed be a more beautiful and striking exemplification of the union of science and art," said Heywood, "than is exhibited in the steam engine." What Heywood saluted specifically was the form to which the steam engine had been brought by the Scottish inventor James Watt.

It was said of Watt that "Every thing became Science in his hands." Born in 1736, from youth he studied and practiced in many fields: woodworking, smithing, metalworking, mathematics, harmonics (he built guitars, fiddles, and organs), chemistry, physics, mechanical engineering. From around 1757, when he began to earn his living as an instrument maker for the physics department of the University of Glasgow, science was in the air he breathed. Later in life he was a member of the renowned Lunar Society of Birmingham, so called because the group—leaders of science and industry, and great minds all—met on nights of the full moon, the better to find their way home after an eve-

For centuries, farmers beat grain with flails and threw it into the air to separate the kernels from the chaff (above). Then, in the mid-1800s, steam-powered equipment rolled onto farms in Europe and the U. S. By the early 1900s, when this picture was taken, a steam-equipped farmer could thresh 13 tons of wheat in one day, about 30 times more than he could have threshed by hand.

ning together. Far more than a brilliant working mechanic as his predecessors in steam power had been, Watt was also a theoretician, and both a creature and creator of his era. It was in this age, the second half of the 18th century, that science and technology fused in the inseparable partnership they still enjoy today.

During the winter of 1764 at the University of Glasgow, the 28-year-old Watt was at work repairing a demonstration model of a Newcomen engine. As he tinkered, Watt began to develop in his mind the refinements by which his genius would soon make the steam engine the greatest source of power on earth. Using the principle of latent heat developed by his friend, the chemist Joseph Black, Watt quantified the efficiency of the Newcomen engine. He calculated that it expended more heat to rewarm the just-cooled cylinder than to inject the steam. Alternately heating and cooling the cylinder wasted time as well as fuel.

Watt fixed the little Newcomen engine and pondered the problem until the spring of 1765. On a Sunday walk, the idea came to him. He needed two cylinders. One, containing the piston, would stay hot all the time. Another, separate but connected by a pipe and valve, would be the condensing chamber, and it would be kept cold. The separate condenser prevented the great loss of steam that Watt had seen in the Newcomen engine, and it operated on a quarter of the fuel.

That was only the beginning. Watt patented the separate condenser in 1769, but he could not afford to develop his engine. He formed a partnership with one of his Lunar Society fellows, a manufacturer and entrepreneur named Matthew Boulton. Temperamentally, the two were opposites, Watt gloomy and pessimistic, Boulton optimistic and energetic. At Boulton's Soho Works, a factory near Birmingham, Watt continually made refinements to what was now a true steam engine: a valve system that introduced steam alternately on both sides of the piston, producing a double power stroke; the exploitation of the expansive property of steam, saving even more fuel; a "sun and planet" gearing system that converted the engine's reciprocal motion to rotary motion so it could drive machinery; "parallel motion" which kept the piston rod moving vertically while driven by an oscillating beam; a centrifugal

In a fanciful watercolor, Richard Trevithick's Catch-me-who-can *steam locomotive chugs around a London square in 1808, carrying passengers at the astonishing speed of ten miles an hour. As a Cornish mining engineer, Trevithick (left) had pioneered the high-pressure steam engine at the same time that Oliver Evans was building one in the U. S. The engines were smaller, cheaper, and more efficient—though more dangerous—than their predecessors. Trevithick built a steam carriage in 1801 and three years later launched the first steam locomotive. It hauled ten tons of iron and seventy men nearly ten miles.*

George & Robert Stephenson

▬▬▬▬▬▬▬▬▬
▬▬▬▬▬▬▬▬▬
▬▬▬▬▬▬▬▬▬
▬▬▬▬▬▬▬▬▬
▬▬▬▬▬▬▬▬▬

The young fellow had risen too far too fast, running the mine's steam engine at age 20 when older men still toiled in the pit. Now he would get his comeuppance. A burly miner, face dark with coal dust and jealousy, dared him to slug it out.

George Stephenson saw that victory would take brains as well as fists. Feinting and ducking, he made the big miner punch thin air and grow tired. Then he moved in, raining blows like ten-pound hammers until the battered bully gave up and shook hands.

It was in England in 1801. The lad had demonstrated his greatest assets: confidence, a refusal to surrender, and an uncanny intuition about the way mechanical things—and even fist-fights—work. Though unschooled in engineering and barely able to read and write, he would one day be called the Father of Railways.

England's factories needed endless loads of coal to fuel the industrial revolution, but no transportation moved faster than a horse. Horses powered the wooden tramway that carried coal past the cottage in northeast England where Stephenson was born in 1781. When the boy got a job at the mine where his father worked, he took apart the pumping engine and put it back together to see what made it go. Years later he invented a lamp that burned underwater, using it to fish at night.

At various mines, he ran the stationary steam engines that hauled up coal. He also experimented: His first "travelling engine" pulled eight wagons totaling thirty tons at four miles an hour. Though not the first locomotive, it used the idea of piping exhaust steam up the chimney to increase the fire's draft and produce more power.

When his son, Robert, began school, the two toiled over lessons together. George felt he could advance only by

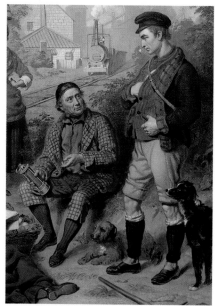

George Stephenson, his son, Robert, and the *Rocket*.

learning engineering—yet he could never retain the simplest theory. He worked by observation and intuition. Robert brought academic science to the partnership they later formed.

Despite his limitations George invented a revolutionary safety lamp that prevented gas explosions in coal mines—and embroiled him in a struggle with the eminent scientist Sir Humphry Davy, who devised a similar lamp and refused to believe that an ill-educated, upstart mechanic from the North Country could have beaten him. So began George's lifelong contempt for establishment scientists.

George next won the position of engineer to build the Stockton & Darlington Railway. Down the track rattled his *Locomotion* as the freight line opened in 1825. It was also the world's first passenger-carrying steam railroad—all of 12 miles long. Already George was dreaming of a nationwide rail web.

The reputations of father and son continued to grow. The two were perfect opposites, yet perfect partners: George sensed the faults in a machine and improved it by trial and error; Robert drew up careful specifications that became models for railway engineers.

The father was a classic self-made man—combative, inflexible when opposed. The son had a gift for managing people but was nervous and insecure, even when famous for designing bridges and railways as far off as Cairo.

With Robert abroad, George took on a daunting project: a railroad to move cotton from the port of Liverpool to the mills of Manchester. The tracks would have to traverse 63 bridges, 2 tunnels, and a vast bog that wouldn't bear a man's weight, much less a train's. Canal companies with friends in Parliament tried to drive out the new rival; so did farmers with shotguns. Steam locomotives, said critics, would kill birds in midair, make women miscarry, fall apart at ten miles an hour, and wipe out anyone nearby. What if a cow got in the way of a locomotive, demanded a skeptic: "Would not that, think you, be a very awkward circumstance?" "Very awkward," George replied in his thick northern accent. "For the coo."

George conquered every obstacle, spanning the bog with a huge raft of brushwood and heather to carry the tracks. Home again, Robert built the *Rocket,* a locomotive with many tubes in the boiler to multiply its power. But doubts lingered: Should steam trains run on the new line—or should it use horses and stationary cable engines?

In 1829 a contest was held. The *Rocket* won, churning past the dazzled judges at 29 miles an hour. A year later throngs lined the route as the Liverpool & Manchester Railway opened with fanfares and a booming cannon. Eight trains carried VIPs—but as one stepped off his train, he was hit by a locomotive on the other track. George sped him to a doctor at a world-record 36 miles an hour. No use; he became the first fatality of the railroad age.

Still, it was the greatest day in railroad history. It set trains on tracks that would thread through almost every land. George Stephenson prophesied: "I will send the locomotive as the great missionary over the world."

Jerry Camarillo Dunn, Jr.

Driven by its inventor, *Locomotion* beats all comers and opens the Stockton & Darlington Railway in 1825.

governor to regulate the engine's speed even with a variable load—the first example of cybernetic control in industry; a gauge that indicated the pressure and volume of steam and thus the power of the operating engine. Watt also gave the world a handy, obvious term to measure the output of work in terms of a technology that the steam engine would challenge—horsepower.

By 1781, according to Boulton, British manufacturers were "steam mill mad." And not only British. Boulton and Watt engines were selling to other countries as well. Matthew Boulton was fond of boasting to prospective clients, such as Catherine the Great of Russia, that he sold what the whole world wanted: power.

The high quality of the Boulton and Watt engine was due partly to improvements made by associates such as the ironmaster John Wilkinson. Cylinders were always a problem. In 1774 Wilkinson called on his experience in boring cannon and produced for the Watt engine, with its monster piston four feet in diameter, a tight-fitting cylinder that prevented the loss of steam. Boulton and Watt enjoyed a patent monopoly until 1800, though they often had to fight infringers in the courts. By then they had built some 500 engines—compared to the more than 1,000 Newcomens and some 30 Saverys at work in Britain. When his patent expired, Watt retired, wealthy and honored, to his country home, where he continued to work on inventions in his attic workshop.

The Boulton and Watt engines altogether had a horsepower of about 7,500. Britain's total steam-produced industrial horsepower at the turn of the century has been estimated at between 20,000 and 30,000. Only 50 years later the figure had risen to 500,000. These were stationary engines, hard at work in industries all over the nation. But to get the steam engine on the move, it took a Cornish mining engineer named Richard Trevithick. Watt said he "deserved hanging."

Robert Stephenson battled nature to build railroads. In the 1830s workers on the London & Birmingham Railway sank deep shafts and linked them for the 7,200-foot-long Kilsby Tunnel (opposite). When quicksand flooded the works, Stephenson installed 13 steam-driven pumps, which siphoned 2,000 gallons of water a minute for 19 months before the tunnel dried out. In the 1840s, for the Chester & Holyhead Railway, he built a tunnel and viaduct at Penmaenmawr (above), altering the design after a gale ruined a 600-foot section of his seawall.

OVERLEAF: Steam engines awaiting assignment line up at South Africa's Bloemfontein Station in 1970, when trains came and went every five minutes. Plentiful coal extended steam's reign here, but by 1976 diesel and electric engines ruled the tracks.

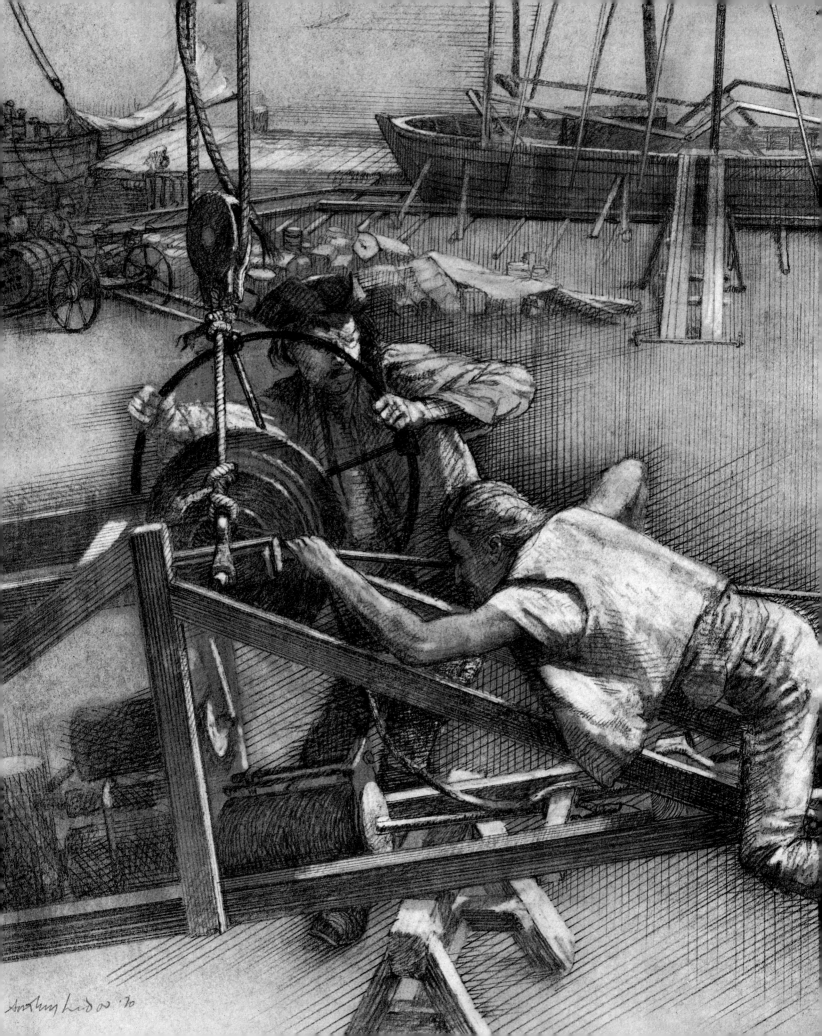

Trevithick had been making model steam locomotives while he waited for Watt's patent to expire. When it did, he wasted no time. It was Trevithick's use of high-pressure steam that concerned James Watt, who knew it could be dangerous. But boilers were being steadily improved. Trevithick developed engines without condensers that blew the steam into the air—and became known as puffers. He used a wheel to transform the action of the piston to rotary motion. Steam engines grew smaller, relatively compact, lighter in weight, and portable.

On Christmas Eve in 1801 Captain Dick's puffer—the first steam carriage to carry passengers—took an abortive tour of the Cornish countryside, defeated by bad roads and a poor production of steam. Two years later Trevithick built a locomotive that traveled on rails and carried a heavy load of workmen and freight at nearly five miles an hour. The trial demonstrated to skeptics that smooth iron wheels could maintain traction on smooth iron tracks. The railroad age had begun.

But the glory was to belong to others than Richard Trevithick, especially to George Stephenson, who made railway travel practical and profitable. Trevithick was called impetuous, restless, violent, and one of the greatest inventors who ever lived. He was into everything but persisted at nothing. In 1816 he left England to install his engines in the Peruvian mines, and for ten years never communicated with his large family. By 1827 he was penniless. In Cartagena he encountered George Stephenson's son, Robert, whom Trevithick had known when Robert was a child in the Stephenson cottage. Robert Stephenson, also in the mining business in South America, paid Trevithick's passage home to England, where the old engineer died a pauper six years later.

George Stephenson's successes with his early locomotives—praised for their fine workmanship—led to the opening, in 1830, of the 30-mile Liverpool & Manchester line, the first public railroad to use steam power to haul passengers as well as freight. Fifteen years later Britain had 2,000 miles of track. By 1852 over 7,500 miles of track carried paying passengers according to strict timetables.

On the North American Continent, with its sweeping distances linked by great rivers, it was the steamboat that first caught on. In 1787 delegates to the Constitutional Convention took time off for promotional rides on the steam-powered, paddle-propelled vessel of John Fitch. Characterizing himself and his partners as "Lord High Admirals of the Delaware," Fitch began a scheduled passenger service between Philadelphia and Trenton that ran smoothly for most of the summer of 1790.

With his assistant at the flywheel, Connecticut Yankee John Fitch labors to adjust the axletree of a seven-ton steam engine in the spring of 1787. Installed in a boat docked at Philadelphia, this engine powered six paddles on each side—a design patterned after Indian war canoes. Fitch promoted his invention for 13 years but never convinced Americans of the steamboat's convenience.

Robert Fulton

On the birth of his son in 1808, Robert Fulton rejoiced, "Every wheel, pinion, screw, bolt, lever and pin about him is of the best proportion size and strength. This has been a fair and successful experiment." It was natural for this man to describe his own flesh and blood in mechanical terms. For Robert Fulton, his work colored everything in life.

Fulton—the man who *didn't* invent the steamboat, but made its use practical and profitable—was born on November 14, 1765. He spent his early years on his family's farm near Lancaster, Pennsylvania. Though he went to school, he probably learned most from the artisans around Lancaster, then flourishing as the largest inland city in the colonies.

The young Fulton, it is said, built a manually powered paddleboat, concocted a Roman candle, made his own lead pencils, and proved himself a superb draftsman. In his mid-teens, in Philadelphia, Fulton was apprenticed to a silversmith, and went on to make a living painting miniature portraits and landscapes. In 1786 he decided to go to England to cultivate his artistic talent.

Fulton achieved modest success in London. Benjamin West, the American painter who headed the Royal Academy, befriended him and painted his portrait. It shows a brooding, handsome man, but it reveals none of the aspirations that tugged at the inventor—aspirations to wealth and fame on the one hand, and to public service on the other. Fulton was ambitious, gracious, and patriotic, but also grandiose, self-pitying, and even self-destructive, often in debt, tirelessly seeking patrons and benefactors.

During his 20 years in Europe, qualities that helped make Fulton a great inventor—boundless energy and an unconventional approach to life—also

Fulton and the brig he blew up to test his "torpedoes."

revealed themselves in the varied company he kept. He fraternized with Young Turks like the utopian Robert Owen, the poet Samuel Coleridge, and naturalist Erasmus Darwin. He lived for a year and a half at the Devonshire manor of his first patron, Viscount William Courtenay, a notorious transvestite. In Paris, he formed a lifelong attachment to diplomat Joel Barlow and his wife Ruth, lived in their household, and at Barlow's urging, traveled widely with Ruth.

Fulton grew increasingly interested in the mechanical arts. He designed new canal systems and then became obsessed with the idealistic if naive belief that submarines could end naval warfare. He went to Paris in 1797, where he built the *Nautilus,* a five-man vessel that could dive to depths of 25 feet. He also championed the use of "torpedoes" that would be maneuvered by auger and line to the hulls of ships.

Neither idea won over the French but while he was in Paris Fulton met a wealthy patron. Robert R. Livingston, the American minister to France who negotiated the Louisiana Purchase, owned a monopoly on the development of steam transportation over the waterways of his native New York State, but of the experimental steamboats built

so far in the United States none had proved practical.

On October 10, 1802, Fulton and Livingston signed a historic pact. Fulton agreed to design a steamboat to travel the 145 miles between New York City and Albany on the Hudson River. Livingston would provide the capital and run political interference. Together they proposed to revolutionize American transport.

Fulton experimented in Paris. On August 9, 1803, he successfully launched a prototype paddle-wheel steamboat on the Seine, achieving 2.9 miles an hour against the current and 4.5 downstream.

Finally, in 1806, after a misbegotten detour to London to promote the submarine idea, Fulton returned to the U. S. In New York City he began construction of a vessel: flat bottomed, 146 feet long and 12 feet wide, powered by a 24-horsepower Boulton and Watt engine, propelled by 15-foot paddle wheels on either side. Fulton dubbed it the *Steamboat.* Spectators derided it as Fulton's Folly.

Fulton, of course, had the last laugh. On August 17, 1807, the *Steamboat,* belching smoke from its coal fires, left its Greenwich Village berth and chugged upriver at over 4.5 miles an hour, negotiating the New York to Albany run in 32 hours (sailing sloops sometimes took four days). A Hudson River Valley farmer likened the spectacle to seeing "the devil going up the river in a sawmill."

In 1808 Fulton married Harriet Livingston, his partner's second cousin. A postnuptial note shows how divided his loyalties would always remain. "The honeymoon & Steamboat," he wrote, "go on charmingly."

Indeed, in its first year of operation, the *Steamboat* earned $16,000. Its success both vindicated and victimized Robert Fulton. He lived to see steamboats on the Mississippi and Ohio Rivers. But rivals and imitators appeared on the scene and Fulton spent much time and money in court, defending his patents and the contro-

Fulton's *Steamboat*, later named *Clermont* by the public.

In a self-portrait, Fulton demonstrates his submarine periscope.

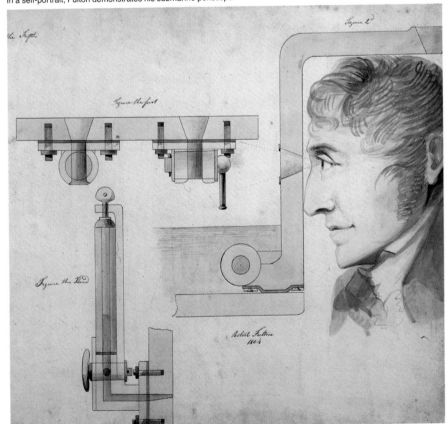

versial Livingston monopoly.

In February 1815, Fulton and his legal team were returning to New York by boat after testifying in Trenton, New Jersey, at yet another lawsuit. As the party trudged across a frozen stretch of the Hudson, Fulton's lawyer fell through the ice, and the other men became soaked while dragging him to safety. Fulton, already exhausted by all the legal wrangling, became ill with pneumonia. Disregarding the fears of Harriet and his doctors, he left his sickbed and returned to Jersey City to supervise the building of his new steam frigate, *Fulton I.* His illness worsened and he died on February 23, 1815.

Robert Fulton recognized that he didn't invent the steamboat, that his achievement was to launch steam travel on water: "As the component parts of all new machines may be said to be old . . . ," he wrote, "the mechanic should sit down among levers, screws, wedges, wheels, etc. like the poet among the letters of the alphabet."

New arrangements bring new ideas.

Stephen S. Hall

His boat used a double-acting engine based on Watt's. But it lost out to the competition of sailboats, which were almost as fast and cheaper, and stagecoaches, which were expensive but much faster.

James Rumsey, a protégé of George Washington, Benjamin Franklin, and other influential citizens, developed a boat on a propulsion system that Franklin had devised as early as 1785. Jets of water that pumped out the stern moved the boat while a steam engine powered the pump. But this was not an efficient source of power. In England, Matthew Boulton offered Rumsey a partnership, as he had James Watt, and the rights to Boulton and Watt engines, but Rumsey's loyalty to his American supporters led him to refuse. In the end Rumsey, like Fitch, failed to build a steamboat that operated reliably and economically.

That was accomplished by Robert Fulton and his partner, Robert Livingston, under the protection of their monopolies in New York and

Steamboats take on passengers and cargo at Memphis, Tennessee, in 1906. With steam-driven paddle wheels, the boats plied the Mississippi River from New Orleans to St. Paul, hauling everything from cotton to cranberries and opening the frontier to settlers. On floating palaces like the Great Republic (above), ladies sat aft, away from the heat and explosive danger of boilers.

Wrapped around massive brake drums, chains with 60-pound links secured the British ocean steamer Great Eastern *before its 1857-58 launch, which took three months and required more than a thousand men. The ship was nearly 700 feet long and five times bigger than anything else afloat. Designer Isambard Kingdom Brunel had sought to build a luxurious vessel capable of carrying 4,000 passengers to Australia without stopping more than once to refuel. But, driven by a screw propeller, a pair of paddles as big as Ferris wheels, and backup sails, the* Great Eastern *handled poorly in squalls and burned far more coal than anticipated. Brunel died soon after the launch, and the ship spent a few years making unprofitable runs to America before being used in 1866 to lay the first transatlantic telegraph cable, from Ireland to Newfoundland (left).*

Louisiana, and after the monopolies were broken, by their competitors. John Stevens was one. When the monopoly prevented Stevens from entering the Hudson River with his *Phoenix,* he sent the boat steaming out New York Bay into the Atlantic and thence to Philadelphia, making it the first steamboat to go to sea. Because of Stevens's efforts, Congress established the U. S. Patent Office, annulled state monopolies, and assumed federal control over American navigation and commerce.

Oliver Evans was another. Evans built high-pressure engines that were compact, cheap, efficient, and eventually dominant on the Mississippi and Ohio. Another rival, Daniel French, pirated Evans's design and the multitube boiler, which John Stevens claimed to have originated, and put them in his *Enterprise.* In 1817 it made the trip upriver from New Orleans to Louisville in twenty-five days. By the 1850s the record had shrunk to just under five days. The 1850s were the era of the paddle-wheeled palaces, decks scrolled with gingerbread, carpeted saloons, music, dancing, fine food and wine, riverboat gamblers, and the education of a cub pilot by the name of Sam Clemens.

Engineers wanted to build steamboats to cross the ocean. In England important developments came together in the *Great Western* and *Great Britain,* ships designed by Isambard Kingdom Brunel, a little giant of a man whose Great Western Railway ran some of the world's fastest trains. The first steamship to provide regular transatlantic service, the *Great Western* was Brunel's extension of his railroad. The *Great Britain* had an iron hull, at 322 feet the longest in the world. The ships carried condensers that distilled fresh water from salt, for the boilers. While paddle wheels drove the *Great Western,* the *Great Britain* had a screw propeller, for more speed and greater fuel economy.

At the turn of the century, a new engine, the steam turbine, sparked the age of the ocean liner, and by the 1930s scores of ships crisscrossed the Atlantic, routinely carrying more than 2,000 passengers apiece. In 1961 the Independence, America, United States, Liberté, Queen Mary, *and* Mauretania *(bottom to top) nosed into New York's Hudson River piers.*

Irish engineer Charles Parsons invented the turbine, bringing speed and efficiency to oceangoing liners. The engine channeled steam through a series of blades mounted in wheels on a rotor shaft to produce electricity or turn a propeller. An engineer on the Queen Elizabeth *cleans the curved blades in one of the ship's four turbines in the mid-1960s (left). Each turbine contains 257 blades. Steam enters through guide blades attached to the underside of the raised cover.*

During World War II the liners, painted battleship gray, carried troops, refugees, prisoners of war, and GI brides. The great ships enjoyed a brief postwar heyday but in the 1960s their age drew to a close.

The size of steam engines increased to meet ever increasing demands for power. By the late 1800s the piston-driven engine had reached its limits. Charles Parsons, an Irish engineer who had grown up in a castle, went to work on the problem. Parsons had spent his childhood in a creative, stimulating, scientific atmosphere. His mother was a pioneer photographer and his father a world famous astronomer who built telescopes (among them the largest reflector in the world) and cast and ground the mirrors in workshops on the estate. Young Charles got lots of hands-on experience in these workshops, and his tutors were the astronomers who worked for his father.

After an intensive apprenticeship in engineering, Parsons developed the first practical steam turbogenerator—still today the way most of the world generates electricity. In a turbine, high-pressure steam spins a series of blades and accelerates as it flows, converting its own energy to energy that can power a generator or, Charles Parsons decided, a ship.

The ship it drove was the *Turbinia*, a sleek vessel, slim as a cigar. In June 1897 the *Turbinia* made an appearance off Portsmouth harbor at a naval review for Queen Victoria's Diamond Jubilee. The little ship dashed from the berth where Parsons had hidden it among the spectators and zoomed past dreadnoughts, merchant ships, and torpedo boats, its turbocharged engine putting out 2,000 horsepower, its top speed 35 knots. It was, until then, the swiftest machine ever to take to water. In its wake, under the power of steam, came the great liners, the *Mauretania,* the *Lusitania,* the *Queens,* and all their grand progeny.

The Age of Electricity

By Stephen S. Hall

I t began, like so many scientific revolutions, with a chance observation. The year was 1780, the place Bologna. An Italian anatomist named Luigi Galvani was studying the effects of static electricity on dissected frogs. One day, beside a machine that generated static electricity, one of Galvani's assistants accidentally brushed the nerve of a frog leg with a metal scalpel, and a spark seemed to jump from the machine through the scalpel. The frog leg twitched violently. Galvani, fascinated by this unexpected effect, discovered that when he connected the frog's spinal cord to its muscle with a bridge made of two dissimilar metals, he could send a lifeless frog leg into convulsions.

What force caused this strange reaction? Galvani, in a paper published in 1791, trumpeted the discovery of animal electricity. Galvani had stumbled upon the phenomenon we now call an electric current, and soon the word "galvanism" denoted the presence of a continuous flow of electricity. But what exactly was flowing? No one really knew. Scientists and inventors spent the next century tinkering with that mysterious current.

Alessandro Giuseppe Antonio Anastasio Volta, born February 18, 1745, in Como, Italy, had serious doubts about the existence of animal electricity even though he, like Galvani, used animals for his early experiments. Volta induced convulsions in a live frog by joining its leg and back externally with a circuit made of dissimilar metals. What he did was create a battery, producing an electric current between two metals separated by a moist conductor—in this case, the frog.

Volta set out to construct a more manageable battery. Moist cardboard was easier to handle than a live frog; disks of silver and zinc replaced the bimetallic bridges. Volta built little sandwiches of silver, cardboard, and zinc, each called a cell or a pair, and stacked the cells to make his famous voltaic piles. The higher the pile, the greater the force generated; Volta likened the experience of touching a pile of 40 or 50 cells to grasping an electric fish.

Volta described his battery in a letter to the Royal Society of London in 1800: "This perpetual motion may appear paradoxical, perhaps inexplicable; but it is nonetheless true and real, and can be touched, as it were, with the hands." Volta did not realize that this inexplicable flow of electricity—current—resulted from the chemical reaction between the two metals and a fluid, or electrolyte.

Scientists began to experiment with continuous current. Sir Humphry Davy applied electricity to the study of chemistry. In 1807 he subjected solutions of two well-known compounds, potash and soda, to an electric current; the compounds decomposed in a process known as electrolysis and yielded two previously unknown chemical elements,

In the 1930s, some 50 years after Thomas Edison's Pearl Street power plant first supplied electricity to lower Manhattan, New York City glowed. From the first battery's spark in 1800 to the power plant of the 1890s, electricity came of age. By the end of the century, electrical energy could be made to produce heat and light and to run machines. The new power source spawned a panoply of applications—from the trolley car to the electric frying pan—and changed forever the way people lived and worked.

which Davy named potassium and sodium. In similar experiments Davy added calcium, magnesium, strontium, barium, boron, and silicon to the list of elements.

Davy installed a 2,000-cell battery in the Royal Institution's basement in London. The device's enormous current allowed him to demonstrate new and wonderful electrical effects. When the terminals of the battery were connected by a wire, the wire grew hot and glowed—became incandescent. When Davy interrupted the wire connection with two pieces of charcoal, then pulled them slowly apart, a large spark jumped across the gap. By holding the charcoal pieces close together, he made a constant spark—an electric arc. Davy had anticipated the invention of both arc and incandescent lighting.

Of all Davy's discoveries, perhaps the greatest was Michael Faraday. Born near London, the son of a blacksmith, Faraday had little formal education. At age 14 he began a seven-year apprenticeship with a London bookbinder and seller; the books he bound kept his fingers and his brain agile—the better to serve him later in his career as an experimentalist. It was the "Electricity" entry in a volume of the *Encyclopaedia Britannica* brought to the shop for repair that drew Faraday to the field he would ultimately revolutionize.

In 1810 Faraday began attending scientific lectures in London. Three years later Davy hired the 21-year-old as a laboratory assistant at the Royal Institution. Faraday spent the rest of his prolific career there.

Young Faraday embarked on electrical research just as the crucial

Italian physicist Alessandro Volta (left) touches a bimetallic loop to an exposed nerve in a pair of severed frog legs, producing enough electricity to make the legs twitch. His audience includes Napoléon Bonaparte (seated second from right), who summoned Volta to Paris in 1801 for a demonstration of the voltaic pile (center right on table). Napoléon was so impressed by Volta's demonstrations

that he made him a count.

Experimenting with a 2,000-cell battery at the Royal Institution in London (above), Humphry Davy produced incandescence and a leaping spark known as an arc—both important to the eventual development of electric light.

Michael Faraday, widely hailed as the "father of electricity," prepares chemicals in his basement laboratory at the Royal Institution. His research led to an understanding of the link between magnetism and electricity, a triumph he recorded in this diary entry of August 29, 1831 (left, upper). Using principles of electromagnetism, Faraday devised the first electric motor and the first generator.

The work of Volta, Faraday, and others sparked a host of experiments in Europe and America. In 1831 a young American teacher, Joseph Henry, greatly increased the power of a magnet by wrapping it in electrified wires, which he insulated with the shredded remnants of one of his wife's silk petticoats (left). The electromagnet would one day provide power for electric motors and generators.

link between electricity and magnetism was established. Hans Christian Ørsted, a professor at the University of Copenhagen, discovered electromagnetism during a classroom demonstration in 1820. He sent continuous current from a battery through a wire and detected magnetism circling the wire: a "very feeble" force, Ørsted noted, but strong enough to deflect the needle of a compass.

Ørsted's experiments caused a sensation in scientific circles. Others, like André-Marie Ampère in France and Georg Simon Ohm in Germany, followed with more work on electromagnetism. In the United States Joseph Henry made large magnets by sending powerful currents through insulated wire wrapped around iron bars.

A decade after Ørsted discovered his "very feeble" force, Faraday found a further link between magnetism and electricity. He started with an iron ring six inches in diameter. He wrapped a coil of wire around half of the ring and a second coil, which he connected to a galvanometer—an instrument that detected the flow of electricity—around the other half. When he attached the first wire to a battery, the galvanometer needle flickered. When he disconnected the battery, the needle twitched again. By starting or stopping a current in the first wire, Faraday had induced, for an instant, a current in the second wire.

Faraday realized that his experiment had made the iron ring into an electromagnet. Could the induced current be a magnetic effect? He wrapped a paper cylinder with copper wire connected to a galvanometer. He moved a simple bar magnet back and forth inside the cylinder, and the galvanometer needle swung along with his motions. This was electromagnetic induction. He used the newly discovered principle in a machine that held a copper disk between the poles of a horseshoe magnet. When the disk was turned, a continuous current ran across it. Faraday had made the first electric generator. All the fundamental pieces were now in place for the development of power-generating dynamos, electric motors, transformers, and electromagnets that would someday lift tons of metal, drive lumbering trams, illuminate entire cities, push sound through a wire.

While Faraday pursued his experiments at the Royal Institution, Joseph Henry was at work as a professor of mathematics and natural philosophy at Albany Academy in New York. Between classes, Henry's students witnessed some of the most breathtaking electrical demonstrations of the era.

Henry began to experiment with electricity and electromagnetism in 1827. He realized that the lifting power of an ordinary magnet could be enormously enhanced with electricity. In perhaps his most persuasive demonstration of electromagnetic power, he bent a 20-inch bar of ordinary soft iron into a horseshoe and wrapped nine 60-foot coils of wire around it. When one coil was connected to a battery the device lifted seven pounds. When all nine coils received current, the same iron bar hoisted as much as 750 pounds. Impressed by these results, the Penfield Iron Works at Crown Point (later Port Henry), New York, bought two of Henry's electromagnets to separate iron ore from rock. This was one of the earliest uses of the new electrical technology in industry.

Another of Henry's experiments, the electromagnetic telegraph, started in 1831 as entertainment for his students at Albany Academy. Henry separated a battery from a small electromagnet with a mile of wire, wrapped all around his classroom. When the circuit was closed, the current activated the electromagnet a mile away. The electromagnet repelled a second, free-swinging magnet, which in turn struck a bell. Whether the electromagnet rang a bell (as Henry's did) or created dots and dashes (as Samuel F. B. Morse's would do), the principle was clear: An electric current could provoke an instantaneous mechanical action at some distant point.

Samuel F. B. Morse, a Yale-educated artist and inventor, thought he could transmit information by "lightning wire." By 1836 he had made a device that sent a message about 50 feet. He would activate and deactivate an electromagnet at the end of a 50-foot wire; the electromagnetic impulses moved a pencil, which marked a moving strip of paper. Less than two years later, Morse was off to Washington, D. C., to apply for patents and search for funding.

While Congress debated the merits of Morse's application, British inventors William Cooke and Charles Wheatstone, on June 12, 1837, patented a galvanic and electromagnetic telegraph. It went into service along 18 miles of the Great Western Railway from London to Slough in 1839, and won public favor soon thereafter when a murderer named John Tawell fled Slough on the London-bound train. His description, sent by telegraph, arrived at London's Paddington Station before Tawell did; he was arrested and later hanged—the first criminal caught by means of electric communication.

Politics won public acceptance of the telegraph in America. After years of struggle, Congress finally approved a $30,000 grant for Morse in March 1843. Morse strung a 41-mile telegraph line with glass insulators from a Washington office in the Supreme Court chambers to the train depot in Baltimore. On May 24, 1844, using his dot-and-dash code, Morse sent his famous first message: "What hath God wrought!" Five days later an answer of sorts came back from Baltimore, where the Democratic National Convention was being held that year. It took nine suspense-filled ballots to nominate a candidate, and so anxious were Washington politicians to learn the outcome that the Senate adjourned and lawmakers crowded into Morse's office to hear, finally, that dark-horse candidate James K. Polk would head the Democratic ticket. From the same office Silas Wright wired to the convention his refusal to accept the vice-presidential nomination.

The impact of the telegraph was global and immediate. On November 14, 1847, market quotations from the London Stock Exchange were telegraphed to Manchester, England, and financial markets would never be the same. In 1849 journalist Paul Julius Reuter instituted a press wire service, and by 1855 telegraphs printed words. The telegraph did its first military duty during the Crimean War. By the end of the Civil War the American telegraph network had grown to cover 200,000 miles. By 1866 a telegraph cable traversed the Atlantic Ocean, and by 1861—eight years before the first transcontinental railroad—

Night becomes day—almost—on the grounds of London's Alexandra Palace in 1880. Steam engines powered the generators, "galvano-electric machines" that provided current to arc lamps. The lamps became a novelty but remained impractical. They were short-lived, expensive to operate, and far too dazzling for household use. "Such a light as this should shine forth only on murders...," said Robert Louis Stevenson.

Arc lights, like those above in New York City, needed constant attention and elaborate scaffolding, even for routine chores such as changing carbon rods inside glass globes.

54

Civilian linemen string telegraph wire for the Union Army near Brandy Station, Virginia, in 1864. Some 16,000 miles of wire—15,000 miles of it controlled by the North—carried military messages during the Civil War.

The man who patented the first U. S. telegraph, Samuel F. B. Morse, first determined to be a painter. He made miniatures, like this 1808 self-portrait, to help pay expenses at Yale. "His aversion to study is unconquerable," despaired a tutor, yet Morse learned about electricity. Studying art in England after graduation, he wrote to his mother: "I wish that in an instant I could communicate the information...." By 1837 he had built the first electromagnetic telegraph; here it stands on the porch of his childhood home in Poughkeepsie, New York (opposite).

the telegraph lines of Western Union stretched coast to coast, putting the Pony Express out of business.

The man who made the leap from telegraph to telephone, not three decades later, was Alexander Graham Bell. "Writing is to me a slow and tedious way of expressing myself," he once told a friend. "I long for one of our old confabulations." The telephone made all manner of confabulation—from business to gossip to courtship—possible at any distance.

A teacher of the deaf, Bell was intimately familiar with the quality of sound vibrations. "The air is always shaking when I make a voice," he said to a deaf student in 1871. Bell began looking for a way to transmit electrically a medley of sound vibrations. Keeping up his rigorous teaching schedule, Bell experimented with what he called the harmonic telegraph. He received financial backing from Thomas Sanders and Gardiner Hubbard, the parents of two of his deaf pupils (Hubbard would later become his father-in-law). And he collaborated with a talented young electrician named Thomas Watson.

During the summer of 1874, while vacationing at the Bell family home in Ontario, Canada, Bell had an insight that inched the telephone a bit closer to reality. He realized that current, rather than remaining

unvaryingly constant, could be modulated to resemble the vibrations of speech. Bell suspected that movement caused by such vibrations on the sensitive mechanism of a mouthpiece could be converted into a modulating current. Experiments in 1875 confirmed this notion.

On the basis of this idea, without a working prototype, Gardiner Hubbard filed a patent application for the telephone on February 14, 1876. As it turned out, Hubbard beat out a rival inventor, Elisha Gray of Chicago, by only a few hours. The patent was issued on March 7, yet it wasn't until three days later that Watson and Bell, working in the attic laboratory of a boardinghouse on Exeter Place in Boston, finally got the device to transmit intelligible human voices. "Mr. Watson, come here! I want to see you," said Bell. Seated in another room, separated by two closed doors and a long hallway, Watson distinctly heard the words through his receiver.

Bell's first telephone bore little resemblance to those found now in every country in the world. It was a box with a hole in the top through which Bell spoke; stretched across the bottom of the box, like the skin of a big drum, was a thin, sensitive membrane called a diaphragm. A human voice made the diaphragm vibrate, and a platinum needle

Thomas Alva Edison

He was a maverick, the scrawny boy with the round face, big blue eyes, and broad brow. Mischievous and inquisitive, six-year-old Tom Edison set the family barn on fire "just to see what it would do" and tried to make a friend fly by feeding him a gas-producing laxative.

After the Edison family moved from Ohio to Port Huron, Michigan, in 1854, seven-year-old Tom was sent to school but couldn't conform to the routine. The dreamy and unruly boy may have been dyslexic; his teacher called him "addled." His mother, a former teacher, began tutoring her youngest son at home and soon had him reading Shakespeare, Dickens, and Gibbon.

At age 12 Thomas Alva Edison launched his business career, hawking newspapers and sundries on the train that ran between his hometown and Detroit. He experimented with chemicals in the baggage car until his makeshift lab caught fire. Between each daily run he read in a Detroit library. "I started with the first book on the bottom shelf and went through the lot, one by one," he boasted later.

In the 1860s Edison roamed the country as a telegraph operator. He often neglected his duties to use the lines for experiments—and once blew up a telegraph station while tinkering with a battery. Hotheaded and stubborn, he never kept a job for long.

But he kept on reading. Michael Faraday's *Experimental Researches in Electricity,* which he read in one day, inspired him to become an inventor. "His explanations were simple," Edison said later of the self-taught British scientist. "He used no mathematics. He was the master experimenter."

With his discovery of Faraday, Edison quit Western Union and devoted all his time to experiments. In 1869 he patented his first invention: an electric

Edison inspects one of his phonograph records in 1916.

vote counter. When he couldn't find a buyer, he formulated a policy he followed the rest of his life: "Anything that won't sell, I don't want to invent."

No scholar, no thinker about nature's deepest secrets, Edison cared little about advancing scientific knowledge. Instead, he became absorbed in making marketable products—and making them quickly.

On Christmas Day in 1871, Edison took time out to marry 16-year-old Mary Stilwell. But not much time. An hour after the service he was back in the lab solving a production problem.

By the mid-1870s the young Midwesterner was well on his way to his reputation as the Wizard of Menlo Park. To this rural New Jersey town, far from the distractions of city life, he recruited scientists, technicians, and machinists to make inventions to order. From his "invention factory" came hundreds of products of collective invention shaped by Edison's vision. He left the technical details to his staff. Unlike most of his fellow inventors, who worked alone, Edison realized the advantages of many hands and minds. With characteristic bravado he promised "a minor invention every ten days and a big thing every six months or so."

Edison ran the Menlo Park lab as

firmly as a medieval abbot ruled a monastery. He drove his devoted "insomnia squads" day and night. In the wee hours he revived weary souls with food and drink, with small talk and music from a large pipe organ in the lab. At midnight organ music boomed from the white clapboard building on the hilltop; flashes of light radiating from the windows illuminated the supernatural aura of the lab and its wizard.

Edison was testing ways to record telegraph messages when he stumbled onto the principles of his most novel invention, the phonograph. Before a few co-workers in November 1877, he shouted "Mary Had a Little Lamb" into this new wonder. The witnesses listened, dumbfounded, as the phonograph needle ran over indented tinfoil wrapped around a metal cylinder and the speaker twanged Edison's high-pitched voice back at them.

Poor hearing helped Edison concentrate on his work but hindered enjoyment of the phonograph, his favorite invention. Deafened by scarlet fever as a child, he hadn't heard a bird sing since he was 12. To listen to the phonograph, he sometimes had to bite into the speaker horn so the sound would vibrate the bones in his head.

Edison came upon many of his inventions while looking for something else. His ability to transform one invention into another—invention by mutation, rather than flashes of inspiration—produced the phonograph as well as the incandescent light, motion-picture projector, and a microphone-like carbon transmitter that improved the telephone's audibility.

In 1878 Edison turned his attention to perfecting a safe and inexpensive electric light to replace oil lamps and gaslights. He was only 31.

Europeans had developed the arc light, but its large, glaring lamps would not do. Electric current leaping between carbon rods emitted noxious fumes from the open globes. If one lamp blew, the rest of the series went out; one switch controlled the entire circuit. To bring electricity into the

A souvenir from Europe, "The Triumph of Light" holds an incandescent bulb above Edison's desk and "spitune."

home of Everyman, Edison envisioned a flameless, glowing filament inside a small, enclosed globe. He would wire his incandescent lamps so each could be turned on and off separately.

With only the vaguest notion of how this could be done, Edison announced with supreme confidence that he'd have the answer in just six weeks. The news created a sensation; stocks of gaslight companies flickered in gloom.

Even though Edison worked himself and his men up to 20 hours a day, the solution wasn't to come for more than a year. They tried dozens of different materials as filaments, including gold, nickel, a fishline, and coconut hair.

In the autumn of 1879, Edison discovered that a charred cotton thread would glow for $13\frac{1}{2}$ hours. A reporter rhapsodized about the wondrous bulb that lit without a match, glowed without flame, its "bright, beautiful light, like the mellow sunset of an Italian autumn." On New Year's Eve a crowd gathered at Menlo Park to see strings of the magical new lights burning brightly in and around Edison's lab.

No system existed to make and distribute electricity to the consumer. So, amid wild public excitement, Edison designed lamps, screw-in sockets, light switches, insulated wire, meters, fuses, even the central power station. When the pioneering Pearl Street station in lower Manhattan came into service in 1882, the world moved from the steam age into the electrical age. The year 1900 saw 24 million bulbs carrying out Edison's promise of "electric light so cheap that only the rich will be able to burn candles."

In 1887 Edison set up a larger invention factory in nearby West Orange. Stocked with "everything from an elephant's hide to the eyeballs of a United States Senator," the lab grew into a complex with 3,600 workers. Over the next four decades he and his assistants tested devices that brought 520 patents, including a motion-picture camera and projector, an improved phonograph and long-playing records, and a business dictating machine.

Edison's notebooks—totaling about 3,500—document the creations of history's most prolific inventor.

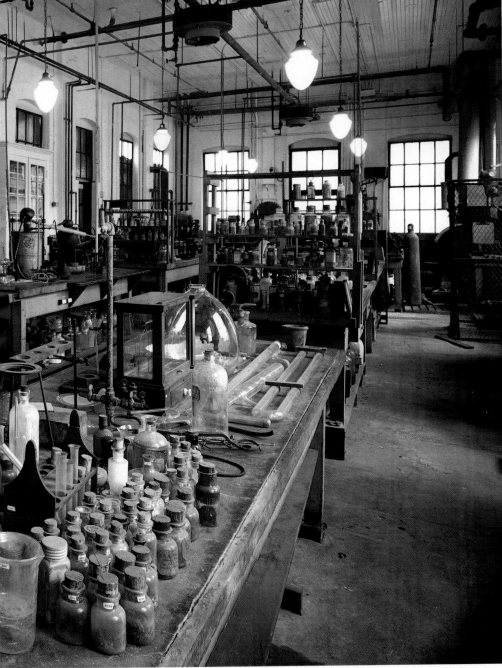

Edison's laboratory in West Orange, New Jersey, prefigured the modern industrial research facility.

Edison catnaps on a 1921 outing with President Warren G. Harding (right) and industrialist Harvey Firestone.

Not all Edison's patents were commercial successes. Fifty thousand experiments resulted in an alkaline storage battery to power an electric car, but Henry Ford's gas buggies swept the market. Edison's inexpensive, prefabricated concrete houses, equipped with concrete furniture, iceboxes, and pianos, never caught on.

Edison could be foolhardy with money. The qualities that led to the light bulb—determination and absolute belief in himself—led also to his loss of General Electric and a $4,000,000 failure to extract iron from low-grade ore with magnets. He ignored household bills. After his first wife died of a brain tumor, he never paid the doctor.

Henry Ford called Edison the world's greatest inventor and worst businessman. Edison's business deals ran from crafty to crooked. He pirated other inventors' ideas and ordered his private eye, Joe "Gumshoe" McCoy, to fix prices and bribe public officials.

As Edison grew older, he grew more bullheaded and coarse. When his second wife reproached him for spitting on the floor, he pointed out that "the floor itself was the surest spittoon because you never missed it." He bathed as little as possible and often slept in his clothes because he believed changing them caused insomnia. He napped anywhere at any time on anything, atop his desk or curled up like a dog on a stack of old papers under the stairs.

Working at night and sleeping by day, Edison hardly saw his family. His wives occupied themselves with parties and pastries, while his six children withdrew into themselves or ran wild.

Edison's scorn for mathematics and "Bulged-headed" theorists (he defined a scientist as "a man who would boil his watch while holding on to an egg") became a passion. He turned his back on progress. He wanted no part of electronics, though his accidental discovery of the principles of the vacuum tube laid the foundation of this new industry. He distrusted alternating current, which Nikola Tesla and George Westinghouse proved could be sent

Col. George Gouraud, Edison's impresario in England, invited musicians to his home for this 1886 recording session.

Early phonographs could record as well as reproduce sound.

An Edison employee delivers wax phonograph cylinders, forerunner of flat long-playing discs.

Typists of the early 1900s transcribe letters recorded on the Ediphone, the first business dictating machine.

farther and cheaper than his direct current could. When New York decided to electrocute criminals, Edison's forces urged officials to use Westinghouse current because it was lethal—and suggested calling this new method of execution "Westinghoused."

Edison continued working into his 80s. It took diabetes, kidney disease, and an ulcer to bring him down on October 18, 1931, exactly 52 years after the birth of his incandescent bulb. He left behind 3.5 million pages of letters and lab notes documenting 1,093 patents, the most ever granted to any individual. Part Bunyan and part Barnum, America's Most Useful Citizen had conceived the electric typewriter, electric locomotive, guided torpedo, fluoroscope, mimeograph, synthetic rubber —and waxed paper.

At dusk on October 21, Edison was buried on a hillside overlooking his West Orange lab. As the mourners—mostly survivors of the age of steam and gaslight—drifted away, the cool glitter of the electric age spread over the valley below. Gone was the wizard who had brought magic to their lives—the magic of recording a human voice, capturing motion on film, and dispelling darkness at a finger's touch.

Carol Bittig Lutyk

62

attached to it moved up and down in a fluid conductor. The louder the sound, the deeper the needle plunged—the less the resistance and the stronger the current. This undulating current produced by the varying sound waves traveled by wire to a receiver, where it reproduced the tones of speech by the interaction of an electromagnet and a strip of metal or reed.

Bell's great triumph occurred when he demonstrated his telephonic apparatus at the Philadelphia Centennial Exhibition in June of 1876. Some 50 scientists and dignitaries took turns at a receiver listening to Bell read one of Hamlet's soliloquies. Bell and Watson presented a series of entertaining lectures to great public acclaim, all the while extending the distance of their conversations. In October 1876 they chatted between Boston and Cambridge, a distance of two miles. By February 1877 Bell, after lecturing in Salem, dictated a reporter's account of his talk over the phone to Watson in Boston; the story appeared in the *Boston Globe* the next morning as "Sent By Telephone."

On July 9, 1877, Hubbard, Sanders, Bell, and Watson founded the Bell Telephone Company based on what has been described as the most lucrative patent in history. Within four years of Bell's Shakespeare recitation, there were 48,000 telephones in use; by 1890, 228,000. But this increase was accompanied by prolonged patent litigation, and soon thereafter Bell dropped his work in telephony in favor of other projects. "I think I can be of far more use as a teacher of the deaf," Bell wrote his wife, Mabel, "than I can ever be as an electrician."

Ironically, a failed telephone experiment by another of Bell's rivals, Thomas Alva Edison, led to the phonograph. In 1877, while toying with a vibrating diaphragm in his laboratory in Menlo Park, New Jersey, Edison impulsively attached a pinpoint to the membrane and yelled into the transmitter. As he hollered, tinfoil passing beneath the vibrating stylus registered the vibrations. Edison believed the squiggly tinfoil indentations would, retraced with a stylus, reproduce the sound. On July 18, 1877, Edison's "Halloo!" became history's first recorded sound.

The nature of electric current had yet to be fully understood by Edison or anyone else, but no one in the 19th century put electric current to broader or better use than Edison.

The path to Edison's electric light was neither easy nor obvious. Humphry Davy had conducted experiments with arc lights at the turn of the century, and improved versions had come into use by the 1850s, but arc lights were so brilliant that their use had to be limited to streetlamps, floodlights, and the like. The first patents on incandescent lights were issued in Britain in the 1840s. As early as the 1860s, Newcastle chemist Joseph Swan created incandescent effects with elements of carbonized paper and cardboard that, heated by the electric current running through them, glowed with such brilliance that they became a source of light. In December of 1878 Swan demonstrated a lamp with a carbon-rod element to the public and received patents on it, but the lamp was never developed commercially.

Two problems made incandescent light impractical. Battery power was too expensive, and lamp elements tended to oxidize or to melt. By

A workman takes a break from laying underground cable for New York's Pearl Street power station. The nation's first generating station opened on September 4, 1882, lighting 800 lamps. By 1895, when artist W. Louis Sontag painted "The Bowery at Night," electricity was a commonplace commodity. The year 1910 saw workers at the Edison Lamp Works in Harrison, New Jersey, testing massproduced incandescent bulbs; their light, according to the New York Times, was "soft, mellow, and grateful to the eye . . . without a particle of flicker and with scarcely any heat to make the head ache."

Nikola Tesla

Nikola Tesla would sit at his table in the Waldorf-Astoria dining room and ritually clean the already pristine silverware and crystal with precisely 18 napkins. He liked his numbers divisible by 3—and you can never be *too* careful about germs.

Tesla's compulsions and phobias were many; he would never touch another person's hair, "except perhaps at the point of a revolver." But at the turn of the century the public recognized in this man with the hypnotic, steel blue eyes the electrical wizardry that set the world alight.

With his invention of an alternating current system that made long-distance power transmission possible, Tesla—not Edison—opened up the age of electric light. He beat Marconi in demonstrating the wireless. And, in 1898, he exhibited radio-controlled model boats and torpedoes. His experiments also anticipated radar, X rays, solar power, and the atom smasher.

Born in 1856 in Smiljan, Croatia, now a part of Yugoslavia, Nikola Tesla was blessed with a photographic memory. He could memorize entire pages of books at a glance. But he was also plagued by apparitions. To escape these periodic flashes of light, in which pictures appeared before his eyes, he began to take imaginary journeys to other worlds. "This I did constantly until I was about seventeen," Tesla recalled, "when my thoughts turned seriously to invention. Then I observed to my delight that I could visualize with the greatest facility. I needed no models, drawings or experiments."

After studying at the University of Prague, he went to work in 1881 for the Budapest telegraph company. It was there, as Tesla was walking through a park with a friend, that a vision of an alternating current motor

Tesla, in 1894, holds a cordless bulb lit by radio waves.

descended upon him. Earlier he had become obsessed with the creation of such a motor. Now his mind's eye had seen not only the motor but also a whole new system for the generation and distribution of electric power.

To commercialize these ideas, he moved to America in 1884. After a stint working for Thomas Edison, who, Tesla felt, cheated him out of a promised bonus for improving Edison dynamos, he formed his own company and won 40 patents for AC equipment. When entrepreneur George Westinghouse bought the patents, the stage was set for a struggle over which system, Tesla's AC or Edison's DC, would power America's factories.

Edison's strategy was to claim publicly that AC was deadly. To prove him wrong, Tesla would invite the press to his Manhattan laboratory for demonstrations in which he picked up lamps and lighted them, without wires, by letting high-frequency current stream through his own body.

Tesla's AC system won out over DC after Westinghouse used it to illuminate the World's Columbian Exposition in Chicago in 1893. A few years later Tesla's generators harnessed the power of Niagara Falls, a feat he had

foreseen in childhood.

Other incidents seemed to show Tesla's supernatural powers. When he envisioned a cloud floating away with angelic figures, he instantly knew that his ill mother had died. Another time he detained his party guests so they missed a train that crashed.

Certainly his intuition helped him divine the secrets of science and apply them to his inventions. Tesla's gas-filled tubes were the forerunners of neon and fluorescent lighting. To generate extremely high voltages, he invented the Tesla coil, important to most radio and TV sets. In 1899 at Colorado Springs he created lightning in a partially successful attempt to transmit electrical energy without wires.

Tesla's brilliance and compelling personality made him a social success among the New York 400. He cultivated the millionaires among them as patrons. From J. P. Morgan he received backing for a "world wireless" plant on Long Island, which he viewed as his crowning achievement. Tesla had conceived of a system that would transmit messages freely—what we now call broadcasting. But a financial panic came along and the backing ended.

After that Tesla's career dimmed. He spent his time feeding pigeons in a park and doted on one bird that was almost pure white. One night she flew through his window as if to convey a message: She was dying. From her eyes came "powerful beams of light," he recalled, "dazzling, blinding light."

When his pigeon died, so did the inventor's conviction that he would finish his projects. On a January morning in 1943 he died in his sleep, at age 87. Already the modern world had started to forget the name of Nikola Tesla.

But in recent years the visionary inventor's memory has been revived. "He was a discoverer of new principles," says his biographer, John J. O'Neill, "opening many new empires of knowlege which even today have been only partly explored."

Jerry Camarillo Dunn, Jr.

A 12-million-volt discharge from the "Tesla coil" lights the Colorado Springs laboratory in 1899.

To Mark Twain, Niagara Falls might have looked like a "beruffled little wet apron hanging out to dry," but to Buffalo, New York, over 20 miles away, the cascade meant a supply of alternating current for electric lights and streetcars in 1896. The tumbling water ran three AC generators, "monsters among electrical machines" invented by Nikola Tesla and built by the Westinghouse Company. Still in

operation, a 5,000-horsepower generator (below) was one of the first three generators to supply practical, inexpensive electric current over long distances. Seven others, installed by 1900 at the Niagara Adams Station Number One, still function too.

the 1870s the first problem was solved. Machines called dynamos converted mechanical energy into electric current. And mercury pumps were being developed to thwart oxidation by sealing the electrical elements within a vacuum. The trickiest problem was finding the right material for the element. Edison experimented with carbon, platinum, iridium, and many other substances.

On the night of October 21, 1879, Edison and his Menlo Park team took a piece of cotton sewing thread and baked it in an oxygen-free environment until it was carbonized. They then connected it to current-carrying wires and inserted it in an evacuated glass lamp. When the current was turned on, it did not immediately burn up, nor did it melt. It simply glowed for $13\frac{1}{2}$ hours. Like any good experimentalist, Edison did not so much celebrate as cerebrate further, immediately seeking more durable and lasting materials. Instead of cotton thread, he tried carbonized bristol cardboard; the bulb burned for 170 hours. He tried carbonized paper, and then bamboo cut from a hand-held fan he found in his office. Carbonized bamboo soon replaced carbonized paper filaments in commercially produced light bulbs.

What really distinguished Edison's lamp from its many competitors

Alexander Graham Bell

The inventor, said Alexander Graham Bell, "can no more help inventing than he can help thinking or breathing." By his own definition, Bell was the perfect example.

His inventions ranged from hydrofoil watercraft to a man-carrying tetrahedral kite to the telephonic probe that searched (without success) for the assassin's bullet that was killing President James A. Garfield. He thought up a top that would cry out when spun and a device that shouted "Help!" when whirled. And he created the audiometer to detect vestigial hearing in the deaf, whose cause became his lifelong preoccupation. By age 30, Bell had revolutionized the transmission of speech and communication itself.

Nature and nurture got Alexander Bell off to an ideal start. The third generation of his family to study the human voice, Bell was born in Edinburgh, Scotland, on March 3, 1847, the middle child and a solemn one, overpowered by his family's ebullience. When he was 11, he added the name Graham, searching for an identity separate from his larger-than-life namesakes, his father and grandfather. The latter, a cobbler turned actor, eventually became an expert on elocution in London. When Aleck was 15 years old his grandfather decked him out in silk top hat and scaled-down cane and had him declaim Shakespearean soliloquies.

Aleck's father, Alexander Melville Bell, was best known for inventing "Visible Speech," a system that taught speech to the deaf by assigning symbols to distinct positions of the tongue, lips, and throat. The interest wasn't only professional; his own wife had become deaf as a girl.

Alexander Melville Bell encouraged his son toward a career in elocution,

Bell demonstrates his telephone in 1892.

and young Aleck obliged, despite—or perhaps because of—a strained relationship between the two. As father had outpaced grandfather, son longed to outdistance father.

Aleck had inherited his mother's musical talent, and he saw himself as a keyboard virtuoso. This talent with musical tones would later come into play when he conceived the telephone. Reflecting both sides of the family, young Graham Bell, as he called himself, took jobs teaching music and speech when he finished high school. Already the teacher-scientist, he delved into the physiology and physics of speech, here obtaining a human ear for study, there translating the talk of a Zulu sailor into Visible Speech. But when his two brothers succumbed to tuberculosis, the elder Bells decided that their surviving son's only hope was the healthier climate of the New World. In 1870 the family reluctantly emigrated to Canada.

Two years later young Bell moved to Boston, the hub of American technology and academia and a bastion of "can-do" capitalism. Here he taught

Visible Speech. But that was not all. He had also been nurturing the notion of a "harmonic telegraph," capable of transmitting multiple messages simultaneously over a single wire. His early experiments with tuning forks had demonstrated the basic concept, sympathetic vibration. He had read his Faraday, Morse, Henry, and Helmholtz. What he needed now were electromagnets, circuitry, receivers, transmitters—and someone much handier than himself to put them all together. The task would fall to a young electrician, Tom Watson, who worked in a shop where Bell bought electrical supplies. Watson deemed himself inadequate to the task: "A dozen young and energetic workmen would have been needed to mechanize all his buzzing ideas," he later mused.

Gardiner Greene Hubbard—Boston Brahmin, patent lawyer, and patron—put it another way when he wrote, in exasperation, of "the tendency of your mind to undertake every new thing that interests you and accomplish nothing of any value." Hubbard's daughter Mabel, deaf since childhood, would become Bell's wife in 1877.

In addition to his experiments, Bell continued to teach and lecture. At times he grew discouraged: "Such a chimerical idea as telegraphing *vocal sounds* would indeed to *most minds* seem scarcely feasible enough to spend time in working over."

March 7, 1876, brought Bell U. S. Patent 174,465 for the telephone, the most lucrative patent ever won. Not mere vocal sounds, but Bell's rendition of "God Save the Queen," his clipped "How do you do?" and his immortal command to Watson crossed a slender steel reed.

The thump of pounding gavels turned the triumph bittersweet over the next few years. Some 600 cases, reflecting the most extensive patent litigation in history, tested Bell and his fledgling telephone company. He won every case, but the ordeal of protracted litigation, along with suggestions that he had purloined his ideas, wounded

Birthplace of the telephone: Bell's Boston attic workshop, now reassembled.

Beinn Bhreagh Hall near Baddeck, Nova Scotia (above). Bell with Grosvenor granddaughters, 1908 (below).

the sensitive young scientist deeply. "I am sick of the telephone," he burst out, "and have done with it altogether, excepting as a plaything to amuse my leisure moments." With the Bell Company's stock valued at $995 a share in 1879, after a deal was struck with rival Western Union, Alexander Graham Bell could afford to amuse himself in his remaining 43 years. And in later life so many new pursuits muted his old obsession that he could remark: "I have become so detached from it that I often wonder if I really did invent the telephone, or was it someone else I had read about?"

He was widely recognized. On a

Bell "chats" with a favorite friend, Helen Keller.

train a conductor implored: "Do you happen to have a telephone about you? The engine has broken down and we are twelve miles from the nearest station." Bell the workingman's friend also made the royal rounds on his far-ranging travels. After a demonstration, Queen Victoria described the telephone as "most extraordinary"; Bell described Her Majesty as "humpy, stumpy, dumpy." The inventor's own appearance was changing to match his public stature. Jet-black hair whitened prematurely; side-whiskers and moustache were overtaken by a great bushy beard; gauntness yielded to girth. Bell was variously likened to Moses, Santa Claus, Apollo, and a polar bear.

Reluctantly donning "cardboard" (a suit), he gave speeches and interviews, joined a multitude of organizations, presided over scientific groups and other worthy enterprises, and became president of the National Geographic Society in Washington, D. C., where he settled with Mabel and their two daughters. Yet he was ill at ease in his public roles: "I often feel like hiding myself away in a corner out of sight," he once lamented. And sometimes, to avoid social engagements, he would lock himself in a bathroom or crouch behind carpet rolls in the attic (where he was smoked out by the smell from

Bell and assistant, Hector McNeil, with a model for a hydrofoil in 1912.

Aircraft engines power the hydrofoil HD-4 across Baddeck Bay in 1919.

his own cigars). Solitude suited him, and he was at his best—working, playing the piano, wandering—while the world slept. "To take night from me is to rob me of *life,*" he wrote Mabel at three o'clock one morning.

Inevitably, Bell's numerous projects and eccentric habits exacted their toll: Intimacy eluded him, especially with his young daughters, who felt the sting of jealousy toward his young protégée, Helen Keller. Bell had longed for a son. After baby boys, born prematurely two years apart, survived only a few hours each, he put his grief to work by inventing the vacuum jacket, a predecessor of the iron lung.

The tragedy of his sons' deaths drew him closer to Mabel, the one person in whom Bell would confide. As Gardiner Hubbard's young daughter, she had secretly thought the frumpy young Scotsman a "quack doctor" over whose

The ring kite, made of tetrahedral-shaped trusses, takes to the air in 1908.

proposal of marriage she waffled; by middle age he would be her "mainspring." In turn, he had fallen for her pluck and charm in the face of deafness; her uncanny talent for lipreading strengthened his claim that the deaf should be taught to speak and read lips, rather than learn sign language. Ironically, she became to Bell "the chief link between myself and the world outside."

More and more, Bell wanted to withdraw to a world of his own creation—Beinn Bhreagh, a sprawling estate on Nova Scotia's Cape Breton Island, where rugged headlands and salt lochs reminded him of Scotland. At first, Beinn Bhreagh was a rustic retreat from Washington's infernal summers, a place where, cigar in hand, he could float on a lake under starlight. Later, the construction of a large house, styled after a French chateau, made Beinn Bhreagh into a gathering place for the increasingly large Bell clan and for the talented young inventors and tinkerers whom Bell encouraged. Over the years, the spacious grounds and surrounding lakes became a great outdoor laboratory for his experiments with hydrofoils, airplanes, kites, and twin-bearing sheep.

On August 2, 1922, in the early morning darkness he liked best, Alexander Graham Bell died at Beinn Bhreagh from the complications of diabetes. It would take years for many of his ideas—solar heat, air conditioning, alcohol-based fuel, to name just a few—to come into general use. As it was, his name was attached to 30 patents, most unrelated to the telephone, yet he usually gave his occupation as "teacher of the deaf."

Leslie Allen

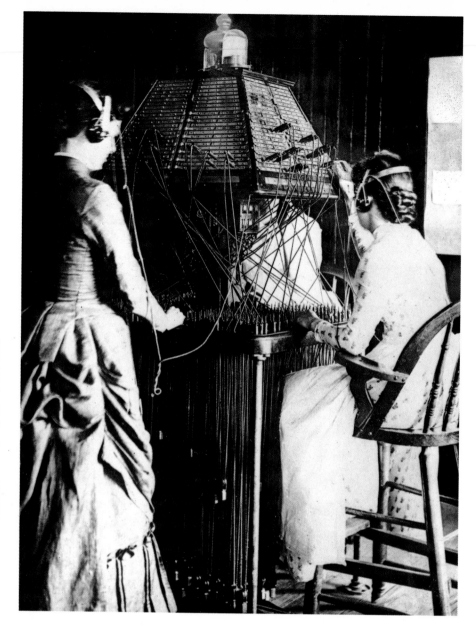

The brothers Harris in San Francisco (above) say a transcontinental good-night to Mom and Dad in 1916. The boys' voices were broadcast at a banquet in Washington, D. C., honoring Alexander Graham Bell 40 years after the telephone was patented. The "lamp-shade" switchboard (above, right) was a late-1800s precursor of the one that carried the Harris's call. Earlier boards were operated by young men but they talked back to customers, so women became the operators of choice.

By-product of a booming industry, outdated New York phone books (opposite) wait to be mashed into pulp.

was Edison's shrewd, overall vision. The electric light by itself, he realized, would exist merely as a technological gewgaw. Edison saw the electric light within a huge system that included power stations and dynamos and distribution networks and fuses and meters and light switches—in short, an entire system for generating, distributing, and utilizing power. Even as he worked on the light, Edison designed the system and the circuitry for delivering power to it. When he finally succeeded, no single invention did more to transform the 19th-century landscape than electric light.

Never averse to attracting the attention of potential investors, Edison located his first commercial power station on Pearl Street, in New York's Wall Street district. When it was time to turn on the system, on September 4, 1882, Edison flipped the switch from the office of his chief financial backer. A single dynamo provided direct current to some

85 subscribers. So habituated was the public to gas lighting that well into the 1880s, wall plaques placed near the new electric lamps warned consumers: "Do not attempt to light with a match."

While Edison refined his power distribution network, an army of electrical engineers worked to improve its component parts. Better dynamos were developed, and the electric motor, first envisioned by Michael Faraday, was now being used for a host of new applications, such as sewing machines, electric fans, elevators.

At this point, simple direct current—from Volta's battery to Edison's lamp—was no longer simple or direct. By the middle of the 19th century, alternating current, so-called because the current changed direction in regular cycles, came into favor. By the 1880s, Europeans Nikola Tesla, Galileo Ferraris, and Michael von Dolivo-Dobrowolski had developed motors that made efficient use of alternating current. There

"Undoubtedly your wife wishes to be considered among the progressive members of her community," advised a 1908 advertisement in Popular Mechanics *for the "ideal vacuum cleaner." She could be, thanks to Murray Spengler, who patented the first widely marketed cleaner in 1908. Spengler and his cousin William B. Hoover founded the Hoover Company, whose name has been associated with housekeeping ever since. By the 1950s a bevy of electrical appliances modernized kitchens.*

began a technological clash between alternating current (AC) and direct current (DC) that came to be known as the Battle of the Currents. Though war was waged on both sides of the Atlantic, the American battle defined the outcome.

The key creative force behind the AC revolution was Tesla, an eccentric and brilliant émigré who worked briefly for Edison in New York before striking out on his own. Given to neurotic behavior and mystical experiences, Tesla claimed that the inspiration for a successful alternating current motor came to him in 1882 in a feverish vision as he walked through a Budapest park. Tesla conceived of a way to make two or three magnetic fields circle around a drive shaft, each slightly out of phase with the others; these rotating, multiphase magnetic fields induced the drive shaft to turn, that is, to become a motor.

In 1887 Tesla applied for the first of several patents on his two-phase and three-phase induction AC motors. As these patents were issued, between 1888 and 1896, Pittsburgh industrialist George Westinghouse bought them and hired Tesla to help his engineers develop a wide-ranging, efficient AC network for commercial and consumer use. Westinghouse and his chief engineer, William Stanley, realized that AC offered a crucial advantage over DC: It could be transmitted at high voltage over considerable distances, then stepped down to lower voltage by transformers for domestic use. As Edison's Pearl Street power station demonstrated, low-voltage DC could be economically transmitted only over short distances. Alternating current looked like a good way to get electric power to a population that was still 75 percent rural.

Thomas Edison fiercely resisted these technological innovations. Edison's investments in the DC system, both financial and emotional, were considerable. By 1886 there were nearly 60 local Edison power companies, all providing DC. So Edison and his allies went to great lengths to discredit the rival AC system.

During the ensuing Battle of the Currents, the Edison forces shamelessly preyed on the public's ignorance of electricity, mounting a campaign with speeches and pamphlets insisting that AC should be outlawed as too dangerous. A pamphlet distributed in 1888 listed people purportedly killed by AC. When the state of New York adopted electrocution as its form of capital punishment, Edison's forces schemed to place a Westinghouse AC generator in Auburn State Prison and then referred to AC as "the executioner's current."

Despite the smear campaign, Westinghouse steadfastly championed the cause of alternating current. Slowly but steadily, he made believers of engineers, financiers and, finally, the public. In 1891 a mine in Telluride, Colorado, was outfitted with America's first commercial AC power transmission system. But Westinghouse's great coup was at the Chicago World's Fair in 1893. By underbidding Edison, he won the lighting contract for the fair and put on a dazzling display of AC's versatility and reliability. More important, he demonstrated the universal system that Edison himself had recognized as crucial. In a technological tour de force that included AC generators, transformers, and rotary converters that changed AC into DC, Westinghouse showed how a

single AC generating plant could deliver power to a variety of potential consumers, from a trolley requiring 500 volts DC to an electric fan requiring 110 volts AC.

The pivotal triumph of AC over DC came during the Niagara Falls project at the close of the century. In 1890 tycoons J. P. Morgan and William K. Vanderbilt formed the Cataract Construction Company with the aim of supplying Buffalo and its 250,000 residents with power generated at Niagara Falls, more than 20 miles away.

Both systems lobbied heavily for the Niagara Falls project, but Morgan and his partners opted for a two-phase AC system in May of 1893, at least in part due to Westinghouse's exhibit at the Chicago exposition. The Niagara power station generated only several thousand horsepower when it went into service in August 1895, but with additional turbines it reached 20,000 and by 1904 was generating 100,000 horsepower. Not only did the AC system work, but ready access to electric power spurred business and helped new industry develop.

The Niagara power station, constructed jointly by Westinghouse and General Electric, the successor to the Edison Company, went into operation roughly one century after Luigi Galvani described the curious current of animal electricity coursing through the nerves of a frog's leg. Although the exact nature of that current eluded the best minds of the 19th century, they were able to study it, harness it, and manipulate it, and by century's end, the mysterious power of electricity was driving heavy machines in factories, providing light to metropolis and farmstead alike, driving trolleys from cities out to newly accessible residential enclaves known as suburbs, powering household appliances like the electric iron and the vacuum cleaner, and communicating information to all corners of the globe.

Almost as an anticlimax, at least to the public and its increasingly electrified households, came news in 1897 of a discovery by British physicist J. J. Thomson. Fourteen years after the invention of the electric fan, several years after the first electric toaster burned the first slice of bread, Thomson identified that mysterious particle of energy whose flow through wires made current. Science called it the electron.

Neon hit Las Vegas in 1946, and now the city's multicolored extravaganzas rival any in the world. Binion's Horseshoe casino sports one of the world's largest neon signs, with eight miles of tubing.

French engineer and chemist Georges Claude invented neon light in 1910, after he discovered that passing electrical current through inert gases in vacuum tubes produced colored light. Glass neon tubes tinted with various phosphorescent pigments eventually produced some 150 hues. Highly efficient, neon light generates almost no heat, and a single tube can burn for more than 30 years.

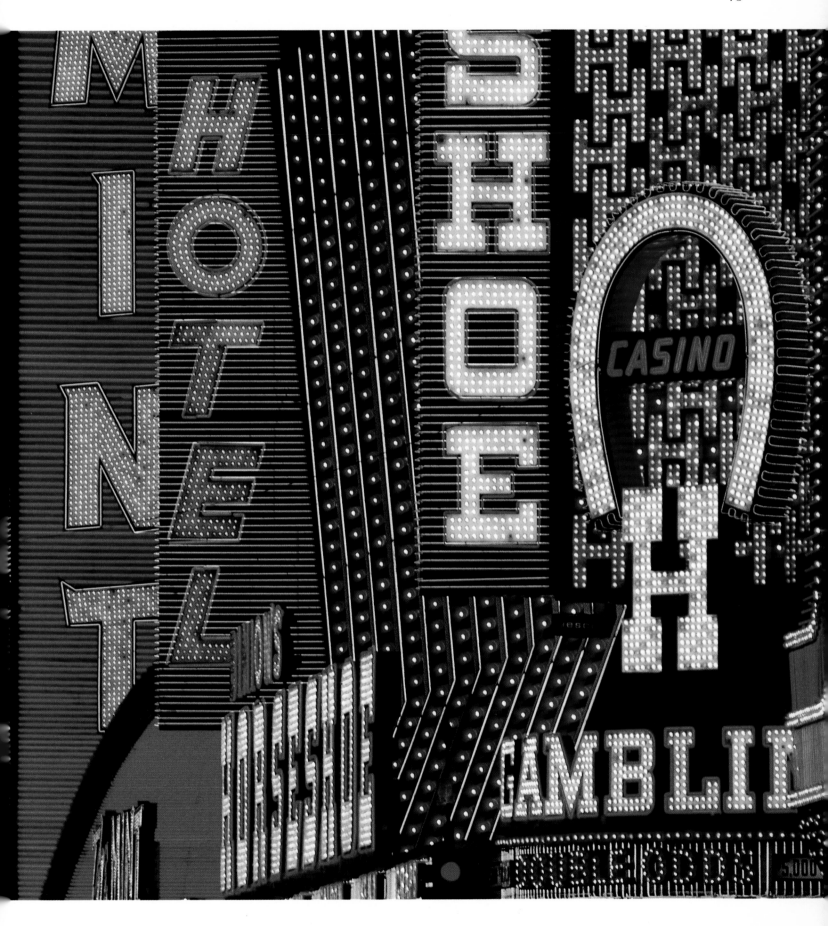

On Wheels & Wings

By Tom D. Crouch

A new kind of vehicle chugged along the streets of Dayton, Ohio, in the summer of 1896. Cordy Ruse, a friend of Wilbur and Orville Wright, had built the first automobile in town. Orville and Cordy spent hours fiddling with the thing; Orville was fascinated by it. Not Wilbur; he said Cordy should fasten a bed sheet beneath the contraption to catch the parts that fell off. And when Orville suggested to Wilbur that the brothers build a "motor wagon" of their own, Wilbur shook his head. Who would ever buy such a thing?

Wilbur Wright was wrong. The automobile was about to reshape American life. It would resculpture our landscape, shrink distances, blur the regional differences between people a continent apart. More than a machine, the automobile would become a social force.

How could anyone in the bicycle trade, let alone someone with Wilbur's vision, overlook the bright promise of the new technology? The bicycle was already paving the way for the coming of the automobile. Cycling had become a national craze. More than a million bicycles were sold in the United States in 1896. Men, women, and children were taking to wheels—and demanding better roads on which to ride.

The average American now glimpsed an alternative to the tyranny of the horse. A young fellow could ride his bicycle swiftly to work six days a week, then pedal into the countryside for a Sunday outing with his best girl. The exhilaration of cycling captivated a generation accustomed to high collars, long skirts, corsets, and the myriad other restraints of polite society. Cyclists went where and when they wanted, under their own power and at their own speed.

The bicycle whetted an appetite that it could never satisfy. Human legs could only pump so fast, and for so long. Farsighted inventors began to envision a huge market for an efficient, reliable "road carriage" that would move under its own power—an automobile.

The ancestors of today's gasoline-powered cars appeared in Germany in 1885. Gottlieb Daimler developed a high-speed gasoline engine and mounted it on a heavy bicycle to create the world's first motorcycle. Carl Benz built a three-wheeled carriage with a front wheel steered by a tiller and his own version of an engine in the rear.

Each inventor combined existing cycle and carriage technology with an internal-combustion engine, an invention that harnessed the power of an explosive vapor—usually gasoline—by igniting it in a closed cylinder. The rapid burning of the vapor pushes a piston, and a crankshaft converts the push to rotary motion that drives a vehicle's wheels. In a sense, the engine runs on a series of explosions.

In 1891 French engineer Emile Levassor built an automobile with the engine up front. Gone was the horseless-carriage shape. In infancy

the automobile had attained its basic form: a four-wheel steerable chassis with rear wheels driven by a gasoline engine ahead of the passengers. That form persists today. Front-mounted engines aligned fore and aft can be bigger and more powerful than engines aligned sideways and mounted under or behind a seat. Capitalizing on that advantage, Levassor in 1895 entered a two-cylinder car in a race from Paris to Bordeaux and back, a daunting distance of more than 700 miles. He finished in less than 49 hours, an amazing feat that caught the public's fancy and helped open a market for the new vehicles.

Charles and Frank Duryea, brothers and bicycle mechanics in Springfield, Massachusetts, read about Benz's experiments and decided—as did many others—to build an automobile of their own. They entered a later model in a competitive trial run in Chicago on Thanksgiving Day, 1895. Hiram Percy Maxim, another early experimenter, described the 11 entries as an "astounding assortment of mechanical monstrosities. . . . every machine needed about five hours of tinkering for every hour of running."

A 54-mile course was laid out on the city streets. Only about half the entrants made it to the starting line, some with rope wrapped around the wheels for traction in the wet snow that had fallen overnight. Maxim rode with Henry Morris in a car with an electric motor. It could run only part of the course on a full battery charge, so Morris planned to stash fresh batteries along the route. But the snow thwarted the plan. "The batteries gave up the ghost," wrote Maxim. "That ended Mr. Morris as far as the race was concerned."

Only two cars finished, one of them the German Benz that Maxim said "looked like a machine shop on wheels." The Duryea averaged about five miles an hour, but it won high honors—and the public's eye.

In 1896 the Duryeas built 13 gasoline-powered cars for sale. An American industry was born. In 1900 sales of all makes reached 4,192 vehicles (a typical price: $1,100); in 1910 sales shot to 181,000.

Even in the early years the consumer faced a variety of options. If he didn't like those cantankerous gasoline engines, he could buy an electric car from the shop of Col. Albert A. Pope, the giant of the American bicycle industry. Pope didn't like gasoline engines either. No sensible

Charles Lindbergh lands near London in 1927 in the plane that bore him over the Atlantic eight days before— and cars bring a crowd to watch. Both car and plane descend partly from the bicycle that put a nation on wheels— and wafted this woman to a parade in the 1890s. The Wright brothers invented the airplane, the Duryea brothers built America's first successful car; all four were bicycle makers.

Gottlieb Daimler &
Carl Benz

The old man lay oblivious to the crowd milling below his window on April 2, 1929. The sad news had spread throughout Germany that Carl Benz, 84 years old, was dying, and motorists had driven in slow file to his house in Ladenburg to pay tribute to the inventor of the first gasoline-powered motor vehicle. In his last 44 years, Benz had seen his brainchild launch a revolution in transportation. On April 4 he died.

Benz had not been alone with his vision in 1885. An unknown rival living only 60 miles away in the German kingdom of Württemberg was wrestling with the same idea. Gottlieb Daimler, ten years Benz's senior, had already patented a small, high-speed, internal-combustion engine by 1883. History gives Benz credit for designing the first true motorcar. But Daimler invented the first practical gasoline-powered engine. Modified forms of that engine still propel most automobiles today. Incredible as it seems, Daimler and Benz never met.

Born in 1834, Daimler grew up in middle-class comfort in the town of Schorndorf, where his family had lived for generations. Steam locomotives hauled passenger trains along Germany's first railroad, built when he was a year and a half old, but people still traveled mostly on horseback or by horse-drawn vehicle. Daimler's father, a baker, had the civil service in mind for young Gottlieb, a model student who excelled in history and mathematics. In 1848, however, war threatened, and Daimler went to work for a gunsmith. Here he acquired metalworking skills and decided on an engineering career.

Warmhearted and fun-loving as well as industrious, Daimler attracted the attention of sponsors, who helped him get a sound technical training that led to a series of jobs in engineering firms.

The Daimler family relaxes on the terrace at Bad Cannstatt: Gottlieb at left, his wife, Emma, at right.

At one of these jobs Daimler met a draftsman, Wilhelm Maybach, who would become his lifelong associate.

From the early 1860s Daimler had mulled over developing a light, compact, fast-running engine—perhaps fueled by a liquid—suitable for propelling vehicles and an alternative to the more cumbersome steam or natural-gas models with stationary fuel sources used in factories. In 1872 he accepted a post with a company that had already patented a new idea for a natural-gas engine designed by Nikolaus Otto. Daimler brought Maybach to work with him and improved on Otto's design, but jealous friction developed between Otto and Daimler, and Otto managed to get Daimler fired.

Retaining his shares in the company, Daimler could afford to move his family in 1882 to a comfortable residence in the resort town of Bad Cannstatt. Here his wife, Emma, hoped to curb his workaholic tendencies. But to no avail. Daimler hired Maybach, who helped him set up a workshop in a greenhouse in the garden. Maybach and Daimler worked such long hours

into the night behind locked doors that the gardener suspected them of counterfeiting and reported his suspicion to the police. Investigating secretly, the police found nothing, and Daimler did not learn of the accusation until later.

Their researches paid off; in 1883 Daimler patented a small gasoline engine fired by a new type of ignition, and in 1885 he mounted one on a clumsy wooden bicycle. Daimler's teenage son Paul, taking the world's first motorcycle on its bone-jarring trial run, would have had a hard time imagining the smooth progress of today's easy riders.

Now Daimler and Maybach experimented with a motor-driven boat. To calm passengers' fears about explosions, they wound wires around porcelain knobs to suggest that the boat ran on electricity. Public apprehension hampered their experiments with a horseless carriage too. Daimler smuggled a reinforced carriage into a factory by night to be fitted with a motor, and tested it on private property and nearby roads in the early morning hours.

Daimler's primary interest was in engines and their ability to propel a

A reproduction of Daimler's 1885 motorcycle stands in his Bad Cannstatt workshop, now a museum.

A 1915 advertisement for a Daimler airplane motor.

variety of vehicles. His inventive genius leapt from boat to carriage to tramway and even to an early dirigible. He did not at first pay as much attention to the design of the vehicles themselves. In this he differed from his younger compatriot, Carl Benz.

Benz's childhood lacked the security of Daimler's upbringing. His father, an engine driver on the railroad, died in 1846 before the boy was two years old. Benz and his mother moved to Karlsruhe, where the widow cooked and took in boarders to pay for her son's education. Her devotion doubtless spurred him on, but he seems to have had a natural bent for science.

Benz, too, acquired a sound training in engineering, followed by practical experience at several firms. In 1872 he set up a machine workshop in Mannheim with a partner whom he later bought out. Poverty dogged the young inventor. He had to sell all his machine tools to pay his debts.

Finally Benz succeeded in patenting a number of parts for gasoline internal-combustion engines, which won him the backing to develop his designs for a self-propelled road vehicle.

Benz's prototype, with its three spidery wheels, resembled a motorized tricycle. The engine lay horizontally in the rear to help stabilize the vehicle. But its electric ignition, water-cooled engine, and differential gear system made it the first gasoline-powered road vehicle designed as an integrated unit.

Early trials were not promising. The car stalled and had to be pushed, wires broke, parts needed replacing. Showing off the car to an audience in the fall of 1885, with his wife, Berta, seated proudly beside him, Benz ran straight into a brick wall. Even many improvements later, he attracted little public interest and few customers.

One day in 1888, Benz's independent-minded wife and her two teenage sons crept out of the house and drove his latest model 50 miles to visit Frau Benz's mother. "Drove" is something of a euphemism. The gears would not change down far enough to handle the hills, so the passengers had to get out and push. Benz corrected this failing by adding a lower gear after the family's tired but triumphant return.

New car designs by Carl Benz and Gottlieb Daimler raced neck and neck into the 1890s. In 1889 Daimler built his first true four-wheel car, with a vertically mounted engine. Benz put out a four-wheel model, the Viktoria, in 1893, and followed it with the Velo—the first car manufactured as a cheap, popular series. Daimler put the engine up front in his 1897 Phoenix. Other companies were starting up too. The age of the automobile swung into gear.

Fear and scorn greeted motoring accidents and breakdowns, and speed limits in England remained at four miles an hour until the mid-1890s. But by 1900 the royalty of Europe was riding in Royal Daimlers or in Benz cars with names like Mylord and Duc, and the craze for car racing had taken hold.

Neither Daimler nor Benz cared much for speed. Their interest lay in improving quality and comfort in low-speed cars for everyday use. But racing

Carl Benz chauffeurs his two-seater Patent-Motorwagen, similar to the three-seater his wife took on an outing in 1888.

enthusiasts, many of them rich customers, publicized new design features and brought in orders. Daimler engines performed with particular élan in a variety of automobile chassis.

One valued Daimler customer, Emil Jellinek, consul general in Nice for the Austro-Hungarian Empire, persuaded the Daimler company to design a new racing model in 1900. He christened it the Mercedes, for his daughter. The triumph of its debut has colored the name ever since. When the rival Daimler and Benz companies decided to resolve their economic differences by merging in 1926, Mercedes-Benz became the trade name for their cars.

Gottlieb Daimler did not live to see the Mercedes launched. Worn out by overwork, he died at age 65 in 1900. It was his colleague, Wilhelm Maybach, who adapted a design of Daimler's son Paul into the model whose significance was memorialized in the words of the secretary-general of the Automobile Club de France in 1901: "We have entered the era of the Mercedes."

Mary B. Dickinson

Mercédès Jellinek, whose name inspired the Mercedes automobile trade name.

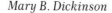

Duplicate of the world's first "automobile," Benz's 1885 three-wheeler, at the Daimler-Benz Museum in Stuttgart.

person, an aide remarked, would take a seat over a device in which hundreds of explosions were taking place every minute. And what about pollution? "Imagine thousands of such vehicles on the streets," growled another critic, "each offering up its column of smell!" Electrics were simple, quiet, reliable—but also slow, heavy, and expensive, and they took a long time to recharge for a short time on the road.

There was a third choice: steam. The Stanley twins, Francis E. and Freelan O., introduced in 1897 the first commercially successful car in America powered by a steam engine—the first of the storied Stanley Steamers. Instead of a roar or a hum, steamers gave a pulsating hiss as they breezed nimbly up to full speed. As early as 1906 a Stanley Steamer whooshed along a Florida beach at better than 127 miles an hour—a speed no human could survive, many had thought. But the lightweight steam engines of the day were too complex for most amateur mechanics to fix. The gasoline engine, it seemed, was simply unbeatable.

And so were American automobiles. Rivals from both sides of the Atlantic crisscrossed the land in endurance runs, kicked up dust in hell-for-leather races, and posed new and shiny at automobile shows. Races served as proving grounds, but the public saw them as little more than dangerous spectacles. It was the endurance runs that convinced the public of the safety and reliability of the American automobile.

Ransom E. Olds was the first American carmaker to succeed at tapping a mass market. From 1901 to 1904 he sold more than 12,000 Merry Oldsmobiles—the famous curved-dash model, a classic horseless carriage with the engine under a buggy seat and the riders sitting high up front, looking as if their horse had just broken loose. But the jaunty little motorized buggy was too light for the needs of most American motorists. It remained for Henry Ford to make a car for the masses.

Wiry and stubborn, Ford was 36 years old when a company was formed in 1899 to build his cars. His backers dropped him in 1902 for wasting time on racing; the Henry Ford Company went on to become Cadillac. Then in 1908 his own firm, the Ford Motor Company, began delivering the most famous automobile ever made—the Model T.

It was a superb machine, a rugged, versatile, high-slung car that won no beauty prizes but could run a rutted farm road without dragging its underbelly. If it did get stuck, you could rock it free with an improved transmission that made shifting between forward and reverse gears fast and easy. Priced at $850 in 1908, the Model T touring car sold in 1916 for $360, thanks to high sales and revolutionary methods

Automobiles crowd Madison Square Garden for New York's first National Automobile Show in 1900. The rich and famous mingled with 48,000 other spectators at the eight-day event. "The handsome gowns of the women and the multitude of snowy masculine shirt fronts," bubbled a reporter, "reminded one of the Horse Show at its best." So did the cars; many looked like carriages waiting for horses.

90

Three Merry Oldsmobiles make a see-saw teeter (below). Dapper drivers jockey the 1901 curved-dash models in a show of the car's responsiveness.

Steam shows its power on the slopes of New Hampshire's Mount Washington (opposite) as a Stanley Steamer chugs toward the summit during a competition held in 1905. Car sales soared as such stunts and contests convinced an eager public of the merits of the automobile.

Three women alight in style from a battery-powered electric runabout in a 1916 poster (lower). Ladies loved the electrics—quiet, smooth-running cars that needed no cranking to start.

of manufacture. "No car under $2,000 offers more," Ford's advertising bragged, "and no car over $2,000 offers more except in trimmings."

Industries had been using interchangeable parts, specialized labor, and an orderly assembly process for well over a century. Ford and his planners forged these and other elements into an integrated mass production system that by 1913 included a moving assembly line. Output topped a million and a half in 1923, and by 1925 half the automobiles in the world were Model Ts. Some 15 million Tin Lizzies had rolled out the doors of the Ford Motor Company by the time production ended in 1927. Convinced that the Model T was the perfect automobile, Ford had failed to keep pace with customer taste. Thanks to him, the public took reliability for granted. But they had come to expect more: a choice of options, appointments, color; a sense of style. The master builder's enormous share of the market shrank as other carmakers—some of

whom were Ford Motor Company alumni—offered the buyer a choice.

In 1912 engineer Charles Kettering developed the self-starter, an electric motor that spared the driver the chore of cranking—and the risk of a broken arm if the engine's compression kicked the handle back. Cranking took muscle; it was considered a man's job. With the starter to do the work, women gained some independence as motorists.

There was a slow improvement in engines as well. Luxury cars of the 1920s and '30s featured engines with eight cylinders in line; more cylinders meant less vibration. If the cylinders were arrayed in two rows angled toward each other in a V, the engine could be shorter and lighter, and that would enable it to accelerate faster to give the car more pep. The first V-8, introduced by Edward Hewitt in 1907, proved too expensive for the early market. But by the '30s an elite few—baronial Packards, stately Franklins, Pierce-Arrows, and Lincolns—cradled long V-12 power plants beneath their outstretched hoods; the regal Marmon even boasted a V-16. Four and six cylinders remained standard for most cars until 1932, when Ford engineers introduced a moderately priced V-8 that quickly became the pacesetter.

Some innovations were little more than advertising gimmicks, desperate efforts to distinguish this year's model from that wonder of

Baker Electrics

Arriving in Style

THE BAKER R & L COMPANY, Cleveland, Ohio

Henry Ford

C rashing noises ripped the night's quiet on Bagley Avenue in Detroit at 2 a.m. on June 4, 1896. In a workshed behind his house, Henry Ford had just finished his first automobile—named the Quadricycle after its four bicycle wheels. But the man who later made a fortune by his obsession with every detail of manufacturing had failed to consider the shed's door. The carriage was trapped, and Ford was demolishing the doorframe and wall with an ax.

Twelve years later Ford created a car so good and so cheap it put America on wheels. After that he revolutionized American industry by perfecting assembly line manufacture. In the process Ford became the nation's first billionaire and arguably its best loved citizen. Whenever people met and the air smelled of gasoline, it was said, there were stories about Henry Ford.

At a time when few celebrities received much mail, 5,000 people wrote to Ford each week. The Ford Craze, as the press dubbed his popularity, nearly put him in the Senate and might have swept him into the White House had he been able to string more than a few sentences together from a podium. But Ford knew that legends sold cars, and with reporters he waxed loquacious. Ford on crime: "Study the history of almost any criminal, and you will find an inveterate cigarette smoker." On charity: "The moment human helpfulness is systematized . . . it becomes a cold and clammy thing." He would tell heart patients in the Henry Ford Hospital to lie on the floor and eat celery instead of heeding their doctors' advice.

Henry Ford bloomed late. Born on a farm in Dearborn, Michigan, on July 30, 1863, Henry was the eldest of William and Mary Ford's six surviving children. He attended school when not milking or sowing. But fiddling with

Henry Ford and his first car, the Quadricycle, in 1896.

greasy gadgets excited him more. When grown he still wrote "much" as "mutch" and thought the War of Independence took place in 1812.

When he was 12, Ford's mother, whom he idolized, died in childbirth. Still obsessed with her memory years later, he ordered the farmyard excavated for pottery shards when restoring his birthplace, then commissioned reproductions of his mother's dishes.

At 16 Henry Ford left home on foot for Detroit, ten miles and a world away. Under the city's pall of smoke, he worked in machine shops for three years, stretching his salary by cleaning and repairing watches at night. He then considered his mechanical apprenticeship complete. Lured by 40 acres his father gave him for the clearing, Ford went back to the farm and married Clara Bryant. Unassuming and devoted, she never gave up darning her husband's socks. She also endured at least one long and public dalliance on his part.

For nine years Henry Ford worked the land and tinkered. Then, on a trip to Detroit, he saw an internal-combus-tion engine. Henry convinced a horrified Clara to pack up and go with him to the city. He yearned to design such an engine to power a tractor but needed to learn more about electricity, on which its firing cycle was based. Ford built his first gasoline engine in 1893, each piece cut from scrap metal. Three years later he built the Quadricycle.

"There was a peculiar sympathy between him and a machine," Ford associate W. J. Cameron would say later. Machines "were living things to him." But the vision that would catapult Ford from mechanic to magnate was slow in coming. Even with backing from some of Detroit's leading citizens, Ford failed twice at manufacturing.

Not until he was 40 years old did Ford hit on the right game plan. "The way to make automobiles," he told an investor in 1903, "is to make one automobile like another automobile, just as . . . one match is like another match when it comes from a match factory." And so he did. That year he scraped together a new set of backers and formed the Ford Motor Company. The fledgling firm sold an impressive 1,700 cars in its first 15 months.

The Model T, born in 1908, made Henry Ford Henry Ford. He identified with farmers, and so made sure it particularly suited rural folk. Rutted dirt roads would not faze it. Its design was so simple that owners could fix the car themselves, sometimes with clothespins or chewing gum. Farmers could detach its engine to saw wood, pump water, run farm machinery. The Model T was Henry Ford in steel, it was said: tough, light, no frills, and a bit unsophisticated on top. The car became sacred to him: Not until 1927, as the 15 millionth Model T rolled off the production line, would he give it up to produce a new model.

"Early to bed and early to rise. Work like hell and advertise," the *Ford Times* advised the company's nationwide chain of dealers in 1911. But the orders poured in so fast that soon Ford became more obsessed with streamlining the manufacturing process than

Workers in Ford's Highland Park factory in Detroit, Michigan, test a Model T motor and chassis in 1914. By then, half the cars sold in the United States were Model Ts.

with his car. The "speed-up king," as his workers came to call him, inaugurated the first moving assembly line in 1913: "Save ten steps a day for each of 12,000 employees, and you will have saved fifty miles of wasted motion and misspent energy," explained Henry Ford. In one day in 1925 a Model T rolled off the line every ten seconds.

As Ford produced more and more cars, his workers complained that they were becoming slaves to machines. By late 1913, Ford had to hire a thousand workers to keep a hundred.

Ford's solution to labor turnover transformed him into an international workingman's hero. In January 1914 he announced his intention to share ten million dollars of profits with Ford Motor Company workers, more than doubling their salary to an unheard-of five dollars a day. The *New York Times* condemned his decision as "distinctly utopian and . . . dead against all experience." But Ford, said company secretary Harold M. Cordell, "didn't give a continental darn" what anybody thought. Henry Ford stood firm. For one thing, he reasoned, his workers could now afford to buy his cars.

The profit-sharing scheme also allowed him to dabble in social engineering. Inspectors from the company's Sociological Department visited workers' homes, inquiring about everything from diet to savings accounts. Henry wanted to make sure that employees were not throwing away his profits on luxuries, liquor, or bought women.

Such paternalism was typical of the man who controlled one of the largest industrial empires in the world. At its height, the gleaming factories were but the empire's flagship. Ford purchased iron and coal mines and a Brazilian rubber plantation; he bought a railway and a fleet of ships to convey materials the last miles to Dearborn.

No one was going to tell Henry Ford what to do. Not stockholders: He bought them all out in 1919. Not business associates: Over the years he drove out anyone with ideas and managerial expertise. And certainly not the

Henry Ford (left) chats with his son, Edsel, on a trip to the countryside in 1921.

At Greenfield Village, Ford greets children outside the one-room schoolhouse he attended.

In a rare quiet moment with his wife, Clara, Henry carves her initials on a tree at Fair Lane, their Dearborn estate.

government: To get around the inheritance tax of 1935, his lawyers created the Ford Foundation to help retain family control of the empire and save it 321 million dollars in taxes.

Ford hated bloodshed, believing that in a previous life he had been killed in the Civil War. In 1915 the Motor King joined antiwar activists on a voyage to war-torn Europe, contributing nearly a half million dollars to the Peace Ship. He promised to burn down his own factories rather than build instruments of destruction.

Two days after President Wilson cut ties with Germany, Ford made a rapid about-face. He dedicated his factories to producing tanks and boats, promising to do so "without one cent of profit." The myth persisted that Ford handed the government the 29 million dollars he gained from wartime production. But he never did.

To further what he considered true American values, Ford purchased the *Dearborn Independent* in 1918. The "Chronicler of Neglected Truth," as Ford called it, praised Prohibition, conservation, and Wilson's plan for the League of Nations. It scorned Wall Street, the gold standard, and monopoly. In 1920 Ford launched a 91-week campaign to "expose" the "International Jew," a figure he blamed for Bolshevism, Darwinism, liquor sales, gambling, short skirts, jazz, and cheap Hollywood movies. Ford's venom won him an award from Adolf Hitler and embroiled him in a lawsuit. He chose to settle the suit out of court and published an apology.

Toward the end of his life, Ford embarked on a crusade to re-ruralize America, as if to atone for his part in its urbanization. To help keep farmers on the land, he devoted thousands of acres in Dearborn to agricultural experiments. During the Depression he grew 300 varieties of soybeans and promoted soy as a new crop for troubled farmers. He sported a silky soy suit and tie, and hosted dinners where all courses—from soup through "coffee" and ice cream—included soybean

Ford once called mass production "the new Messiah." Right: A single day's output in 1915—a thousand chassis.

The Model T, called "the right car at the right time at the right price."

products. He even made car parts of soy-based plastic.

Ford created a 240-acre time machine in Dearborn to teach people about their roots. At Greenfield Village he assembled, piece by piece, the homesteads and workshops of his heroes: Abraham Lincoln, the Wright brothers, Noah Webster, a favorite teacher. He shipped in Thomas Edison's Menlo Park laboratory, complete with carloads of red New Jersey soil. And just next door, Henry Ford dedicated an enormous museum to the very machines that destroyed the bucolic America he yearned to preserve.

Even as Ford puttered in bean fields and old houses, he grew increasingly meanspirited with age. He hired thugs to control workers with truncheon and fist. Union organizers were fired, spied on, beaten up. At his right hand was Harry Bennett, a pistol-toting ex-boxer with underworld connections. "Harry gets things done in a hurry," Ford liked to say.

The company he launched was one casualty of the dictator's whims and stubbornness: By 1946 it was losing an estimated ten million dollars a month. Edsel Ford, Henry and Clara's only child, was another. Although Edsel officially became president in 1918, the elder Ford refused to relinquish power. Believing Edsel too soft to manage a large company, he thwarted every initiative his talented, loyal son ventured, and humiliated him in public in a misguided effort to toughen him. People said that when Edsel died, at the age of 49 in 1943, what really killed him was a broken heart.

The Ford women finally ousted a furious Henry in favor of his grandson, Henry II. Edsel's widow, Eleanor Clay Ford—with full backing from Clara—used the ultimate threat: She would sell her stock outside the family if he refused to abdicate. Fuming, Henry I resigned in 1945. In 1947 he died.

"I invented the modern age," Henry Ford once said. As if in reply, Will Rogers commented about Ford: "It will take a hundred years to tell whether he helped us or hurt us, but he certainly didn't leave us like he found us."

Carole Douglis

Many on wheels but few behind horses, New York drivers fill Fifth Avenue with cars and a two-tier bus in the early 1900s. Some urbanites hailed the trend from horse to motor; for one thing, it cut the stink of droppings.

Motorists weary of prying flat tires off rims welcomed the demountable rim (right), available in 1904. Tires still blew out even on short trips, but now milady could change to a spare— as one tire maker boasted—"without the use of a crowbar or pickaxe."

Many steps ahead of older models, a Rickenbacker shows off its four-wheel brakes in 1924 (opposite) as a Los Angeles crowd shows off its confidence.

technology marketed only nine months before. Other improvements—the automatic transmission, for example, which was first offered in 1934—eventually became standard equipment.

Automobile manufacturing had become the centerpiece of American industry. The automobile had given new life to the fledgling petroleum industry at a time when its chief product, kerosene lamp oil, was being replaced by electricity.

By the 1930s the entire American population could have been loaded into its cars and taken for a Sunday drive. The transformation of the landscape was well under way. Gas stations seemed to spring up on every corner. Roads stretched into turnpikes. Hamlets grew into suburbs and bedroom towns for city commuters. Parking meters sprouted at the curbs, and billboards and drive-in movies adorned the roadsides.

The exuberance of America in the 1950s was expressed in cars agleam with chrome, punctured with portholes, finned and two-toned and whitewalled and powered for performance far beyond what the speed limits would allow. A few new companies tried to break into the market with innovative ideas: Tucker with a steerable headlight, Amphicar with a little convertible you could drive right into a lake and float to the other side. Chrysler tested a turbine engine in which the blast from burning kerosene whooshed across an array of winglike turbine blades to set them spinning, but myriad problems—high cost, slow acceleration, low fuel economy—doomed the innovative engine. Behind the glitter and the valiant failures, cars changed mostly in refinements.

In 1949 an odd-looking little car rolled onto the American road—the German Volkswagen. Unlike big American cars designed to go out of date in a year, the Beetle looked and ran about the same two decades later. By 1972 more than four million had been sold in this country. American automakers were forced to take notice. The heyday of the big, fast, chrome-encrusted automobile was coming to an end.

An oil crisis highlighted the economy of small cars. City gridlock, suburban traffic jams, and air pollution called for a fresh look at our overpowered gas-guzzlers. Consumer advocates pointed to 50,000 traffic deaths a year and questioned the safety of cars big and small.

Lawmakers responded with seat belt laws and emission standards. Planners worked on mass transit systems and set up lanes for car pools. Inventors tinkered—and still do—with ideas new and old: cars that run on solar cells, batteries, steam, and fuels from gasohol to propane.

Despite advances in systems and performance, this year's model is in many ways the younger brother to the clanking smoke maker of a century ago. In infancy the automobile evolved its basic configuration, and before long it had achieved all the speed a road driver would really need. So the modern car is the result of slow and steady refinement rather than dramatic technological breakthroughs. From Henry Ford to Lee Iacocca, the heroic figures of the auto industry have been not so much inventors and scientists as entrepreneurs and managers.

The automobile gave people the power to travel on the ground as never before. That helped rekindle an ancient dream: to travel *above* the ground, not in balloons at the whim of the wind but on controllable

wings that would whisk people where they willed. By the 1890s daring birdmen had flown in gliders; could the new gasoline engine propel such craft as it did the bicycle and carriage? Was a powered, piloted, fixed-wing flying machine—an airplane—now possible? Almost. More than the automobile, the airplane required true invention, the discovery and refinement of principles and hardware unknown before.

From the beginning the goals of those who design flying machines have been diverse, for the airplane has been seen not only as a family conveyance but as a working craft, and especially a tool of war. Warplanes must fly higher, faster, and farther than yesterday; cargo craft must lift more freight; airliners must carry more passengers with greater speed and safety. From first flight to countdown, steady progress has been punctuated by great leaps, breakthroughs by the brilliant and persistent minds of inventors.

Ancient Greeks had imagined Daedalus and Icarus flapping aloft on wings of feathers and wax. Leonardo da Vinci had sketched aircraft with birdlike wings of wood and fabric. The idea persisted that, if humans were to fly, they would probably have to do it like birds—with wings that flapped by the muscle power of the pilot. It was not until the 19th century that English baronet Sir George Cayley built and flew gliders with rigid wings, some of them large enough to carry a man. Cayley was first to argue that a flying machine should have separate systems for propulsion, lift, and control, an idea that foreshadowed the modern airplane.

The year 1896 was a banner year for the flying machine. German engineer Otto Lilienthal had caught the world's imagination with some 2,000 flights in gliders. His success inspired other glider experimenters, notably Percy Pilcher in Britain and Octave Chanute in the United States. That autumn, Samuel Pierpont Langley, then secretary of the Smithsonian Institution, flew a steam-powered unmanned aircraft he named *Aerodrome* for nearly a mile.

Optimism soon turned to doubt. By 1899 both Lilienthal and Pilcher were dead, victims of glider crashes. Chanute and others had lapsed into inactivity. It was an inauspicious time for Wilbur and Orville Wright to enter the field. Yet it was Lilienthal's crash in 1896 that challenged the Wrights to devise a craft controlled not by the pilot shifting his weight, as Lilienthal had done, but by—what?

The creation of a system for controlling the way a flying machine moves through the air was the first of those mental broad jumps that make the history of aeronautics so different from the story of the motorcar. The Wrights, builders of bicycles, were especially well suited to make the leap. They knew that, unlike a motorist, a cyclist has to maintain control in two directions of motion, balancing from side to side by

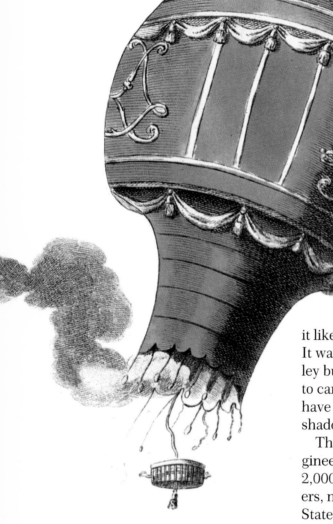

Both man and beast made flights before the Wrights. In 1783 French papermaker Joseph Montgolfier and his brother Étienne sent aloft from Versailles this painted fabric hot-air balloon with flight's first passengers: a sheep, a duck, and a rooster. Louis XVI, Marie-Antoinette, and a horde of subjects watched agog as the craft floated for eight minutes and landed two miles away. Soon humans dared to ascend, and for nearly a century ballooning was the only way to fly.

German pioneer Otto Lilienthal glides down a hill in 1896 (opposite). His 2,000 flights helped lead the way to winged-and-powered flight.

shifts in body weight while steering right or left with the handlebars.

But a craft in the air cannot be steered by its wheels, and weight shifting is a risky way to balance. The Wrights set out to do both by mechanical means. A spiral twist along a wing, they reasoned, would raise one wingtip and lower the other for balance or to bank into a turn.

They built a kite to test the idea. It worked. The kite was the first link in an evolutionary chain of machines with which the Wrights solved the basic problems of controlled flight: one kite (1899), three manned gliders (1900, 1901, and 1902), and three powered airplanes (1903, 1904, and 1905), each a biplane with one wing above the other.

Earlier experimenters reasoned that a wing's cross section—an airfoil—should be convex on top and concave beneath. When such a wing slices through air, its shape produces increased air pressure on the underside and reduced pressure on the upper side. The resulting upward

Wilbur & Orville Wright

e wanted a barrel of oysters. The stranger was from North Carolina's Outer Banks, so the story goes, and he'd been sent over to this Norfolk restaurant to buy some. When asked why, he is said to have explained: "There are two loony Yankees down at Kitty Hawk trying to learn to fly, and they want to eat some lynnhaven oysters before they die."

Orville and Wilbur Wright, the two "loony Yankees" from Dayton, Ohio, had already learned to glide. So had others; flight's pioneers had swooped down hillsides on rickety wings, or sent powered models aloft while they watched from the ground. Tinkerers and headline seekers had climbed into outlandish contraptions to lurch, flap, bounce, and fail in the quest—often comic, sometimes tragic—for powered flight. But never had man, engine, and wing left the ground together under the pilot's control and flown not because of gravity but in spite of it.

The Wrights, too, had swooped— and bounced—in their series of flying devices: a kite controlled by cables, then piloted gliders, and now what they hoped would fly as a powered, piloted airplane. In December of 1903 they hurried to test their ungainly craft and give wings to mankind.

They seemed an unlikely pair for so momentous a role: tall, pensive Wilbur, born near Millville, Indiana, in 1867; dapper, mustachioed Orville, born four years later in Dayton; respectable business partners, purveyors of bicycles, sons of a clergyman and his wife. But in 1878 their father had given the boys an odd little toy of bamboo and paper with two propellers. Wind up the rubber band, and it skittered into the air like a helicopter. The lads were fascinated. The seed was planted.

They wore the thing out, then built several of their own. The small ones

Orville (left) and Wilbur Wright at a 1910 aviation meet.

flew; the big ones floundered. Why?

All their lives Orville and Wilbur asked why of anything mechanical. Once Orville bought a new car and promptly took it apart. His housekeeper made do with an old-fashioned icebox; she wouldn't allow a refrigerator in the house. He'd just take it apart.

Of the Wright children—four boys and sister Katharine—these two were uncommonly close. Yet they argued for hours, said a niece. " 'Tis,' would say Wilbur. 'Tis not,' Orville would argue . . . and often they'd switch ideas in the middle of the argument."

Neither held a high school diploma, yet both read avidly. When their mother died and their elder brothers moved away, the remaining Wrights drew close. Katharine was friend and guiding hand to Orville and Wilbur, neither of whom ever wed. As Wilbur once explained, they could not support "a wife and a flying machine too."

For awhile Orville and Wilbur ran a printshop, then opened a bicycle shop to ride the boom in biking. Here they learned things useful in contriving an airplane: how to transmit power by a chain and sprockets; how to make precision parts; how to control a bicycle not just around its vertical axis, as in steering a wagon, but also around its

Wilbur Wright in 1902, in the shed near Kitty Hawk where the brothers lived, worked, and stored their 1901 glider.

fore-and-aft axis lest it fall on its side.

There was more to flying than to bicycling. An airborne craft has a third axis, along the wings. Nose up, the craft climbs; nose down, it dives. Much was already known about wings and engines, so the Wrights began with the remaining problem: control.

They studied buzzards by the hour, staring at the soaring birds and arguing about how they kept their balance. Birds stay level by twisting their wingtips, but a pivoting wingtip on a spindly glider would be too fragile. One day in the bicycle shop, Wilbur idly twisted a long cardboard inner-tube box—and suddenly saw the solution in his hands: wing-warping. Twist the whole "box" of wings and struts, and the craft would bank.

The glider they built in 1899 showed that wing-warping would work. But in their man-carrying glider of 1902, they found that a bank could start a spiral into the ground. So they devised movable rudders that kept the turn under control. The buzzard watchers had missed one detail: The birds were using their tails to control their turns.

Much of what the Wrights read about aeronautics was wrong. To test their own wing designs, they mounted sections of wings on a bicycle. Encouraged by the results, they built a wind tunnel—and came up with data so accurate that modern instruments can add only minor refinements. As their ideas took shape in kites and gliders, the Wrights sought out the winds and slopes of Kitty Hawk for flight tests. On December 17, 1903, they were ready to fly the world's first airplane: the Wright *Flyer*.

No newsmen came to see it fly. Few people thought it would. Three days earlier, it hadn't. A coin toss had given Wilbur the honor of flying first, but he had overcontrolled, slammed into the ground, and damaged the forward control surfaces. Now it was Orville's turn.

Only four men and a boy came to watch. John Daniels recalled the brothers shaking hands "like two folks parting who weren't sure they'd ever

Wilbur watches Orville pilot their *Flyer*'s first flight, December 17, 1903. Below: Orville watches Wilbur brief sister Katharine for a flight in France in 1909.

news item." Another newsman remarked that it was nice the boys would be home for Christmas.

The telegram even garbled the facts, turning 59 seconds into 57—and history's first airplane pilot into "Orevelle."

No wonder newsmen were skeptical. The eminent scientist Samuel P. Langley had watched his man-carrying *Aerodrome* splash into the Potomac River a few months earlier. Nine days before the Wrights' success, Langley's craft splashed again—"like a handful of mortar," said a reporter. If a scientist of Langley's stature could fail twice—and so spectacularly—then what could a couple of obscure bicycle makers do?

The brothers were as methodical after their *Flyer* flew as they were before. Their agenda: Refine the invention into a practical flying machine, protect it with patents, then sell airplanes. In 1904 they built a second airplane and in 1905 a third, each a result of flight tests, ground experiments, and hard-won experience with the one before.

The new machines flew out of Huffman Prairie near Dayton. Here the Wrights could test and refine until their patent was secure. Twice the local press came out to see if there really was a story here—and twice the flights were scrubbed as the brothers shook their heads and shrugged over engine trouble. Some said they timed their flights between runs of the trolleys that clattered past the field, so that the passengers would see an odd-looking crate on the ground, maybe, but nothing strange zooming around in the air.

A patent was issued in 1906, and the Wrights' new company prospered. But in 1912 Wilbur came down with typhoid fever and died on May 30. A spark in Orville died with him. The company was sold three years later.

History's first airplane pilot lived until 1948. By then Orville had witnessed immense leaps in the science he and Wilbur lifted off the ground—a science then poised for the moon, now for the planets, and tomorrow for the stars.

David F. Robinson

see each other again." Wilbur coached the onlookers not to look glum. Shout, clap, cheer Orville on! Then, on the windswept sands, they watched the fitful birth of the airplane as the *Flyer* rose off its launch rail and bored into a stiff headwind for a 12-second flight before jolting onto the sand. Wilbur ran alongside to steady the wing, so Orville had recruited Daniels to click the camera—and catch the most famous photograph in aviation's album.

Three more flights stretched the *Flyer*'s range to 852 feet, its endurance to nearly a minute. But it was never to fly again. As the brothers stood by it, planning the next flights, a gust hurled the *Flyer* wing over wing along the sand. The men scrambled after it; Daniels took an unscheduled ride when he got tangled in the tumbling spiderweb of wires and spars and fabric. The plane was a wreck, fit only for dismantling and shipping back to Dayton. But no matter. It had flown!

"Success four flights," said a telegram to their home. "Average speed through air thirty one miles longest 57 seconds," it exulted. "Inform Press," it asked, then added, "home Christmas."

The press was unimpressed. "Fifty-seven seconds, hey?" said one Dayton newsman. "If it had been fifty-seven minutes then it might have been a

force is called lift, and it is what raises a plane off the ground. But early data on airfoils was inaccurate, so the Wrights had to redesign their wings, testing and refining airfoils to produce more lift.

How to move a wing forward so it slices through the air? Engines with crude propellers had been tried, but the Wrights reasoned that a good propeller should be an airfoil, a spinning wing that "flew" through the air to create forward "lift," or thrust. Their airplanes flew with two propellers they laminated and carved themselves, elegant airfoils linked to a single engine by chains and sprockets as on a bicycle.

They flew their first powered machine near Kitty Hawk, North Carolina, on December 17, 1903. The best of the day's four flights, 852 feet in 59 seconds, was not very impressive to newsmen of the day, and few papers took note of the triumph. The brothers went home to Dayton, where they flew all but unnoticed for two more years. By the end of 1905 they had achieved their goal: a practical flying machine that could stay aloft for an extended time under the full control of the pilot.

The Wrights captured the honor of inventing the airplane for themselves and their nation, but America's lead did not last long. News of their machine inspired a flurry of aeronautical effort in France. On July 25, 1909, French flyer Louis Blériot startled Europe with a flight across the English Channel in a monoplane, a single-wing craft such as Cayley had envisioned, with rudder and elevator at the rear of the fuselage. In this design, the airplane, much like the automobile, achieved in infancy the form still expressed—in supersonic fighter and space shuttle.

The airplane was still a frail thing of wood, wire, and fabric, but the implications of Blériot's feat were enormous. War clouds were gathering and, as one writer warned, "England is no longer an island." No country could let a potential enemy forge ahead in this new technology. Governments set up flying schools and research programs; newspapers sponsored contests to encourage innovation. In a scant decade, Orville and Wilbur's brainchild was ready for war.

Now the public found a new kind of hero. Newspapers and magazines portrayed airmen as inheritors of chivalric tradition, daring young knights who had escaped the filth and horror of the trenches to meet the foe in single combat in the sky. Many fought without parachutes, radios, or armor plating; some flew sitting on iron stove lids so at least their manhood would be shielded from bullets from below.

The margin of victory in the air was often slim—a bit more speed, a tighter turn, a slightly faster climb. Each side rushed new aircraft

As aviation matured, a nation found heroes in the sky. Crowds everywhere ringed Charles Lindbergh, as here on St. Thomas in 1928. The Lone Eagle checks his engine's rocker arms—as always, leaving nothing to chance.

Fame eluded most other veterans of the airmail runs. In open cockpits, unsung stalwarts like William Hopson (left) huddled in heavy flight suits, defying uncharted terrain, fin-

icky engines, high death rates—which soon included Hopson himself—and myriad other hazards to fly the mail. A Swallow biplane (upper) re-creates airmail's shaky beginnings in a run from Washington to Idaho in 1976.

OVERLEAF: Pilot's-eye view of a 1930s DC-3. Gyroscopic instruments like those at center helped keep these airliners on course in low visibility.

through design and production and into battle. Most still held to the biplane design; some had three wings, like the famed Fokker triplane that bore Germany's legendary Red Baron to 21 of his 80 kills and finally to a last dogfight over the pastures of France.

The Great War ended, but aviation still gave an adoring public its heroes. Barnstormers, wing walkers, and record breakers awed small-town fairgoers and made big-city headlines. Another grail beckoned the brave: the $25,000 Orteig Prize for the first nonstop flight between New York and Paris. In 1927 Ryan Airlines and a lanky barnstormer named Charles Lindbergh teamed up to build a plane designed for a single flight. While competitors readied teams of crewmen, some with multiengine planes, Lindbergh proposed to solo the Atlantic and do it on one engine. The *Spirit of St. Louis* was a flying gas tank. It had a radial engine, with cylinders jutting like daisy petals to be cooled by the airstream instead of by a radiator system heavy with water. The savings in weight helped make the 3,600-mile flight possible. There was now no challenge, it seemed, that plane and pilot could not master.

Buoyed by Lindbergh's achievement, engineers began to design a new generation of aircraft. Better airfoils, improved propellers, retractable wheels, and drag-reducing engine cowlings appeared in the 1920s and '30s. Wider use of aluminum alloys allowed new approaches to structural design. The cantilever wing, which supports itself without external wires or braces, came into common use.

The best of the new machines, the Lockheed Vega, took shape on the drawing board of John Knudsen Northrop. The son of a Santa Barbara contractor, Jack Northrop was 20 in 1916 when he stepped into a local garage that housed the Loughead Aircraft Manufacturing Company. Already a draftsman and auto mechanic, he was in love with flying and eager to try aircraft design. The Loughead brothers, Allan and Malcolm, gave Northrop his chance.

First flown in 1927, the Vega was a high-wing monoplane built partly of plywood with an enclosed cabin and extraordinarily clean lines. "It was some beautiful airplane," said an admirer. "There wasn't anything like it in the air." The sleek "plywood bullets" from Lockheed Aircraft—now spelled as its founders' name sounded—seemed to fit everyone's notion of what an airplane ought to look like.

Some of America's most famous pilots earned their fame flying across the continents, over the oceans, and around the world in souped-up Vegas. Amelia Earhart soloed a Vega over both Atlantic and Pacific waters, set a transcontinental speed record in 1932—then broke her own record a year later. Wiley Post, a one-eyed Oklahoma wildcatter, flew his *Winnie Mae* twice around the world. Later he would fit a supercharger to the engine, help engineers design a pressure suit for himself, and climb into the cold, thin air of the substratosphere to explore the great air current called the jet stream.

Flying was definitely still for heroes. Fledgling airlines wooed the courageous traveler into what one writer called "an ordeal of endurance and discomfort, mingled with occasional moments of stark terror. Only the brave, or the extremely impatient, flew the airlines. . . ." For the

Crowds at New York's La Guardia Airport watch passengers board a Douglas DC-3 in 1940. Air travel came of age with the DC-3. Wings set low on the fuselage allowed wing spars to pass under the floor without obstructing the cabin. When pilots added power, variable-pitch propellers would pivot to take a bigger "bite" of the air, thus giving more thrust while the engine ran efficiently at a constant speed. Hundreds of DC-3s still fly; one has logged 84,000 hours.

Luxury came of age too. Planes called flying boats lured fares from the ocean liners with fine cuisine, attentive stewards, and portholes (left).

brave or impatient, a typical airliner of the 1920s was the Tin Goose, the sway-bellied Ford Tri-motor with its corrugated aluminum-alloy skin and starter buttons any Model T owner would recognize. Inside the cabin, the passengers—about a dozen—found stand-up headroom, wicker seats, and some of the industry's first flight attendants. But airline routes were limited and airliners were noisy and slow; the Tin Goose cruised at about 110 miles an hour. The time was right for a new generation of airliners.

From Jack Northrop's drawing board came advanced construction methods, embodied in sturdy, streamlined aircraft like his Northrop Alpha and Gamma. Such airplanes presaged the classic Douglas DC-3, a dependable, versatile, twin-engined workhorse that revolutionized air transport before World War II. Hundreds of DC-3s still carry freight and passengers half a century after the Gooney Bird took shape in the

VOUGHT - SIKORSKY
VS·300
STRATFORD   CONNECTICUT

Inventor Igor Sikorsky eases his VS-300 helicopter off the ground on an early test in 1940. The Russian-born immigrant was his own test pilot for a series of designs, the first a failure in 1909. Tethers were later discarded as he taught himself to manage a rotary-wing craft—not an easy task, for a helicopter is inherently unstable. Each rotor blade is a wing, whirling through the air to generate lift even as the craft hovers motionless. To move in any direction, the rotor blades must change pitch as they whirl. Also, the spin of the rotor in one direction tends to spin the fuselage in the other. Tail rotors help control this torque.

On demonstration flights, Sikorsky hovered his craft and flew it forward, backward, up, down, and sideways. By 1941 the inventor and his VS-300 had set a helicopter endurance record: about an hour and a half in the air.

mid-1930s and at last persuaded the traveling public to go by air.

Again the war clouds gathered. Again the airplane, a tool of commerce that was reshaping old ideas of time and distance, was seen as a weapon—and again it would win or lose on the strength of its speed and agility. But engineers had long known there were limits to how high up a piston engine could operate and how fast a propeller could move an airplane through the sky. The best planes of the late 1930s were already operating close to those limits. Another great leap was needed.

Two engineers, one English, one German, made the leap. Frank Whittle was a 22-year-old pilot at the RAF Central Flying School in 1929 when he conceived a new breed of engine: the turbojet. Air would be forced in the front through a fanlike compressor, mixed with fuel, and ignited. The exhaust would drive a turbine at the rear that powered the compressor, and the rearward blast would push the engine forward.

English inventor Frank Whittle, between two aides, works on his jet engine in the 1930s. Though war loomed, he saw his Whittle Unit as an engine for mail planes: "I wasn't thinking of war." But the Germans were. As Whittle experimented, so did Hans von Ohain, on an engine much the same. In 1939 von Ohain's drove a Heinkel He 178 on the first flight by a turbojet. Not until 1941 did Whittle's jet engine fly. In 1944 the German Messerschmitt Me 262 (left) roared aloft as the world's first operational jet—stymied by design and production problems, as were Britain's jets, until too late to alter the war's course.

Whittle developed the idea on his own until he won approval to spend six hours a week, no more, on the project. In April 1937 a prototype roared to life on a test stand—and threatened to blow up as it sped out of control. That, recalled Whittle, "did not do my nervous system any good at all." His system—and his concept—fared better in 1939, when a redesigned engine ran for 20 minutes and convinced officials of its worth.

Unknown to Whittle, Hans von Ohain had tested a jet engine for German plane maker Ernst Heinkel two months before Whittle's 1937 trial run. Von Ohain's engine burned hydrogen; later he worked out the problems of liquid fuel that Whittle was already solving. History's first turbojet airplane, the Heinkel He 178, took to the air at Rostock, Germany, on August 27, 1939. Heinkel stood by the runway, elated. "The hideous wail of the engine," he wrote, "was music to our ears."

Frank Whittle's great moment arrived almost two years later, when his engine first took the Gloster E28/39 aloft. One of the excited spectators recalled exclaiming, "Frank, it flies!" Said Whittle: "That was what it was bloody-well designed to do, wasn't it?"

By then the airplane had redefined war, raining havoc on an enemy's homeland far behind the lines where soldiers clashed. World War II began amid attacks from the sky—Warsaw in Europe, Pearl Harbor in the Pacific—and ended with the destruction of two Japanese cities by two airplanes armed with two bombs, one for Hiroshima, one for Nagasaki.

Not until near war's end did jet planes enter combat as the Messerschmitt Me 262 darted into Allied bomber formations in a last-ditch attempt to guard the German heartland. Then, in the final days, a stubby wonder streaked skyward on a column of smoke—the Komet, a German interceptor that hurtled aloft on a rocket engine. Jet and rocket came too late to turn the tide of war, but they profoundly influenced the decades of peace, when millions would jet around the global neighborhood and talk to each other via satellites rocketed into orbit.

Rockets have been fired for centuries, mostly as festive fireworks. The principle is simplicity itself: Fuel is burned in a chamber, the exhaust is concentrated through a nozzle at the rear, and the reaction to the rearward blast propels the device forward. Adapted to aircraft and spacecraft, such engines can achieve phenomenal speeds. Thus it was that flight's next threshold was crossed by a rocket plane.

In a dive a fighter of World War II could almost approach the speed at which sound waves travel through the air. But it would encounter severe buffeting and control problems. What if it went supersonic? Was there a "sound barrier" that no plane could penetrate without breaking up or tumbling out of control?

On October 14, 1947, Capt. Charles Yeager of the U. S. Air Force flew the Bell XS-1 rocket plane faster than sound—and found not a barrier but a door to the future. Scientists knew a bullet can travel faster than the sound of the shot, and the XS-1 was shaped like a winged bullet. But wind tunnel tests showed how, at such speeds, shock waves form behind the wings to slow the plane down. NASA engineer Richard Whitcomb reasoned that a bullet has no wings; thus a fuselage should pinch into a Coke-bottle shape to compensate for such projections.

Questing minds like Whitcomb's have borne mankind from Kitty Hawk to the moon in a lifetime. Yet, while some look to the planets, others look back to the basics that challenged the first to fly. Like Lilienthal, they ride the winds in hang gliders, refined into a swept-wing shape of synthetic fabric and aluminum tubing, yet flown like gliders of a century ago by a pilot shifting his weight. Like the Wrights, tinkerers fitted such craft with engines—at first chain-saw motors controlled by a throttle in the pilot's teeth, and now elegant little engines bolted to spidery mini-planes called ultralights.

In December 1986 designer Burt Rutan watched his brother, Dick, and copilot Jeana Yeager take off from California's Mojave Desert and head westward in *Voyager,* a spindly craft made of space-age composites with one wing ahead of the other. Nine days later the strange airplane circled in from the east. Two brothers working on a shoestring had led a team that created the first airplane to circle the globe nonstop.

While big-budget research and development programs push the limits of human flight, there is still room for the maverick inventor with a good idea. Where will that questing human spirit carry our children, our grandchildren? For them, as for us, not even the sky is the limit.

NASA engineer Richard Whitcomb (above) checks a wind-tunnel model built in the 1950s to test his idea that a pinched-waist fuselage could ease a plane to supersonic speeds with less drag and buffeting by shock waves. Impatient with lengthy redesigning, Whitcomb would test a model, then reshape it with files and putty until it tested right. His concepts shaped a generation of supersonic jets.

An X-15 rocket plane drops away from its mother ship in 1962 (right) to roar to 246,700 feet—nearly 47 miles up. Hurled to the fringes of space by the brute force of its liquid oxygen-liquid ammonia engine, the X-15 could streak a mile in less than a second, more than twice a bullet's speed—and 146 times the speed of the Wright Flyer *that got aviation off the ground a scant six decades before.*

A World of New Materials

By Robert Friedel

n the spring of 1851, a large and remarkable building took shape on the lawns of London's Hyde Park, a short ride from the royal and parliamentary palaces of Westminster. Even a casual observer would have suspected that the glass-and-iron structure was designed by a builder of greenhouses, but this time Joseph Paxton's building had a very different purpose: It was to house the first great world's fair. In this Crystal Palace, almost 14,000 exhibitors, half of them foreigners, would display the greatest works of craftsmanship, industry, and technology to more than six million visitors. For Americans in particular, the exhibition offered an opportunity to display the products of their ingenuity to scoffing Europeans. For one American especially, it was the setting of a momentary triumph in a life beset with difficulty and tragedy.

Fifty-year-old Charles Goodyear came to London full of hope and optimism; he also came laden with the fruits of almost two decades of labor—dozens of articles of India rubber. Goodyear put on a remarkable show: The walls, roof, carpets, and furniture of his Vulcanite Court were all rubber. On the rubber-veneered bureaus and cases were rubber combs, buttons, canes, and musical instruments, rubber toys and air mattresses. And over it all hovered a giant rubber raft and rubber balloons; some were six feet in diameter, others were painted like globes of the world. The exhibition jury was impressed, and so, apparently, was the public. For many this was their first view of a new material—a substance that behaved like nothing anyone had ever seen.

Today we are accustomed to the novelty that so impressed the Crystal Palace's visitors. New materials, with wonderful and useful properties, are part of the expected order of things. A miracle fabric in one's shirt, a new alloy for a jet engine, a high-tech reinforced plastic for the latest tennis racket, or a ceramic that goes right from freezer to oven—these are accepted parts of the promise of modern science and technology. In the middle of the last century, however, even after the marvels of the steam engine, the telegraph, and hundreds of other inventions had helped make a world "mad for improvement," novel materials were almost unknown. This situation was about to change radically.

Charles Goodyear, who had found a way to transform rubber from a hard gum into a stable, moldable, marketable material, died in July 1860. Nine months later the American Civil War broke out. The Union armies consumed great amounts of rubber—27 million dollars worth by war's end in 1865—and in so doing launched the American rubber industry. At first its products consisted mainly of combs, buttons, erasers, blankets, conveyor belts, fire hoses, and waterproof clothing.

Rubber gained importance with the 1888 invention of a Scottish-born veterinarian. John Dunlop had a successful practice in Belfast,

Ireland; he also had a bit of the tinkerer about him. To improve the ride on his son's tricycle, he substituted air-filled tires for the customary solid rubber ones. Impressed by the results, Dunlop patented his invention and organized a company. His timing could not have been better, for his pneumatic tires rode the crest of the bicycle craze of the 1890s. The new tire also made the ride easier on automobiles as thousands of horseless carriages took to the road. In the 20th century the pneumatic tire converted rubber from a material of convenience into a strategic commodity, one of the handful of materials that soldiers fought and died for. During World War I the Allies hastened Germany's defeat by blockading its supplies of Far Eastern rubber. In an about-face almost 30 years later, Japan's conquest of Southeast Asia nearly crippled the Allies until American rubber companies, universities, and research institutes banded together to produce a synthetic substitute.

The real materials battles of the new century, however, would not be fought over rubber plantations but in chemical laboratories. One of the six million visitors who attended the Crystal Palace Exhibition in 1851 was undoubtedly a transplanted German chemist named August Wilhelm von Hofmann. He had arrived in London six years earlier to teach at the new Royal College of Chemistry and serve as its first director. Hofmann would have felt obliged to go to the Crystal Palace because the man who had recruited him, Queen Victoria's German-born consort, Prince Albert, was the exhibition's major patron.

But chemical novelties such as Goodyear's rubber would have been enticement enough, for Hofmann had studied with one of the most creative chemists the world had seen, Justus von Liebig. A professor for almost 30 years at the University of Giessen, Liebig had transformed German scientific education by setting up the first laboratory to teach chemical research methods. His work laid much of the foundation of modern organic chemistry. Liebig probably let go of his star pupil with some reluctance, for at 27 Hofmann had already staked out a promising specialty, the constitution of coal tar.

While the vile, smelly by-product of the distillation of coal to coke would hardly seem a substance with a future, in the hands of Hofmann and a few of his students it turned into the stuff of dreams. Over the next few decades it became the source of new colors, odors, tastes, and remedies, either imitating nature's best efforts or, sometimes, surpassing them. In his first publication in 1843, Hofmann had revealed that a key ingredient of coal tar was aniline. One of his English students,

Bright! New! And synthetic. The glistening stainless steel shell of a 1930s diner strengthens the streamlined image of machine age efficiency and cleanliness. Inside, Formica protects counters and tabletops, Naugahyde covers the booths, chrome wraps the counter stools. Food cooked in aluminum pots and pans is served on melamine dishes and eaten with stainless steel tableware.

Since the industrial revolution, new materials—stainless steel and synthetic dyes, plastics and Pyrex—have changed the way we dress, cook, eat, play, work, and get around. Today we live in a world more of our own making than of nature's.

Charles Goodyear

You could say that Charles Goodyear invented the rubber check. The slight, sickly man had a knack for invention surpassed only by a talent for financial disaster. He learned how to transform a malodorous gum from South American trees into vulcanized rubber—a material for waterproofing raincoats, insulating electric wires, and making everything from washers for kitchen sinks to mile-long conveyor belts for industry. But even his success in 1839 didn't end years of frustration and racking poverty.

Typically, Goodyear had begun his experiments in debtors' prison, where he was such a frequent guest that he referred to it as his "hotel." In the kitchen of a small cottage on the prison grounds, he blended raw rubber with anything he could get his hands on: ink, witch hazel, cream cheese, soup, castor oil. By testing everything in the world, he was bound to find the right material sooner or later. His obsession was to solve the sticky problems that gum elastic caused manufacturers. Angry customers were returning rubber goods that, in summer's heat, melted into foul-smelling, gooey masses and, in winter, grew as hard as planks and often cracked.

Born a Connecticut Yankee in 1800, Goodyear had entered his father's farm implement business and later moved to Philadelphia, where he opened America's first retail hardware store. It soon went bankrupt. Then he decided to become an inventor, obtaining patents on a number of tools.

Gum elastic had captured Goodyear's interest during boyhood, when he had found a rubber bottle and marveled at its "wonderful and mysterious properties." As a young inventor, he visited New York and tried to interest the Roxbury India Rubber Company store in his design for an improved life preserver valve. The manager glumly told him that what the company really needed was improved rubber.

Goodyear saw his life's challenge: God had chosen him to improve rubber as a gift to the world. It did not faze the inventor that he had absolutely no background in chemistry, nor any funding for his work.

After several years of experimenting —and those stints in debtors' prison— Goodyear found that nitric acid seemed to divest the raw material of its adhesiveness and enable it to resist heat. This success finally attracted a backer. Goodyear's rubber factory failed, however, when a financial panic hit in 1837. The unlucky inventor camped his wife and children at the derelict factory on Staten Island and fished for their food in the harbor.

Finally, Goodyear obtained a contract with the U. S. Post Office for 150 rubber mailbags, to be treated with nitric acid and sulfur. He confidently stored the bags in a warm room while he and his family were away. When he returned, the bags lay decomposing in a mass on the floor. Apparently the process cured only the surface of the rubber. His work had yielded nothing but more heartache and poverty.

Then came the breakthrough.

While doing experiments in 1839 at a Massachusetts rubber factory, Goodyear accidentally dropped a lump of rubber mixed with sulfur on a hot stove. To his shock the lump didn't melt, but charred like leather. When he nailed it outside in the cold, it remained flexible. He had discovered vulcanization, the process that was to make rubber a commercial success.

By now exhausted and ill, Goodyear sought financial backing to perfect the process. But he had prematurely trumpeted his "successes" once too often, and no one was interested. Fearful that he would die before he could convince the world, he pawned the family dishes and sold his children's schoolbooks to keep his research going. His family once again stood by him.

Goodyear experimented in his kitchen lab for more than four years before discovering a way to keep rubber from getting sticky in summer, brittle in winter.

Objects from Goodyear's era include a book with rubber pages as well as a rubber fan and jackknife. Goodyear gave his wife the watch, chain, key, and seal—all made of rubber and set with gold, diamonds, and rubies.

invented, but one day their tires would lead rubber to glory.) He felt he had discovered a "vegetable leather" or an "elastic metal" suitable for any use.

The process of vulcanization was too easy to duplicate, however, and soon patent pirates fell upon him. Although Goodyear was eventually successful in court, the costs ate up all his profits. The aging barrister Daniel Webster defended the patent, charging $15,000 for two days' work in court, more than Goodyear had earned in half a century.

On a promotional trip to Europe, Goodyear spent 30,000 mostly borrowed dollars on a pavilion at London's 1851 Crystal Palace Exhibition. Everything in lavish Vulcanite Court was made of rubber, from carpeting to musical instruments. Millions of people saw it, but there was no financial triumph. Partly because of his delays, Goodyear failed to win patents in both England and France. When Napoléon III awarded him the prestigious cross of the Legion of Honor for an exhibition in Paris, the medal had to be delivered to prison, where the inventor was locked up for his latest debts.

Goodyear's career had taken more bounces than a rubber ball—or a rubber check. As an acquaintance once described him: "If you meet a man who has on an India rubber cap, stock, coat, vest, and shoes, with an India rubber money purse without a cent of money in it, that is he."

The inventor died in 1860 in a New York hotel, en route to New Haven for the funeral of his favorite daughter. It is likely that lead, an ingredient in his experiments, finally poisoned him. He left enormous debts, but eventually his estate made his family comfortable. Although rubber products brought wealth to others, none to himself, the inventor always remained philosophical: "The advantages of a career in life should not be estimated exclusively by the standard of dollars and cents," he wrote. "Man has just cause for regret only when he sows and no one reaps."

Jerry Camarillo Dunn, Jr.

After several more years of experimenting, Goodyear determined that a usable rubber resulted from applying steam at 270° Fahrenheit for four to six hours. Varying the amount of sulfur in the mixture made the rubber harder or softer. Now he could apply for a patent.

But again Goodyear snatched defeat from the jaws of victory. He gave away licenses for rubber products at ridiculously low royalties and decided to forgo manufacturing to dream up other uses for his material. In his notebook he jotted ideas for inflatable life rafts, sails, rubber bands, baptismal pants, self-inflating beds, and wheelbarrow tires. (Automobiles had not yet been

A worker in a B. F. Goodrich rubber factory vulcanizes hot-water bottles in 1939.

Charles Mansfield, showed how to fractionate coal tar into dozens of chemicals, including aniline, benzene, and toluene. Some years later, in 1856, Hofmann hinted to another student, 18-year-old William Henry Perkin, that other more complex but useful products might be synthesized from these simple chemicals. The professor suggested that Perkin try to produce the antimalarial drug quinine.

Perkin could not synthesize quinine, but what he did discover founded an industry. Starting with aniline, he carried out a series of reactions to build up a larger, more complicated molecule. The result was not white quinine crystals but a black powder. What Perkin did next probably owed more to youthful exuberance than to native genius. Rather than throwing out the powder, he mixed it into boiling water. The liquid turned brilliant purple. On another hunch, he dipped in silk strips and watched them soak up the purple color. He dried them and then tried to wash the color out. It didn't fade. After a week out in the sun, the strips were still bright. Perkin had discovered the first coal-tar dye.

The discovery of aniline purple, or mauve as the enthusiastic French called it, spawned the great synthetic chemical industry. Perkin's success in manufacturing synthetic dye in England infected chemists throughout Europe with a mania for dyes. They manipulated coal-tar chemicals to produce not only startling colors like magenta and imperial purple, but perfumes and flavorings as well.

No one was better prepared to take advantage of the new possibilities than the Germans. Their recently unified nation built up its industrial

Colorful and colorfast, the first commercially successful synthetic dyes (opposite) evolved from the coal tar that clogged the flues of gasworks. In 1856 Englishman William Henry Perkin (above) tried to synthesize quinine from the sticky black sludge but by accident produced a bright purple solution. Soon mauve was coloring cotton, wool, and silk—and giving its name to the 1860s: the Mauve Decade.

Though created in England, the new industry flourished in Germany, where research laboratories developed synthetic indigo (above left) and other hues. Discoveries in organic chemistry stimulated industrial and military growth, turning Germany into a world power. Ironically, khaki dye from Germany colored the uniforms of British soldiers in World War I.

What began as a shortage of elephant tusks in the 1860s ushered in the age of modern plastics. Seeking a $10,000 prize for an ivory substitute for billiard balls, John Wesley Hyatt, a printer from Albany, New York, concocted celluloid. This tough, pliable semisynthetic plastic led the way from natural plastics, like amber and shellac, to modern synthetics.

Celluloid took shape as buckles, brushes, combs, game pieces (below), jewel boxes and gemstone-inlaid jewelry (opposite), and dozens of other elegant, mass-produced consumer goods. As wipe-clean "linen," celluloid collars and cuffs (right) became an affordable emblem of upward mobility for working-class Americans of the late 1800s. But celluloid was highly flammable. And celluloid dentures softened in hot water, literally curling the teeth of tea drinkers. The solution: less flammable synthetic plastics, notably Bakelite. Celluloid, though, still garners its share of the sports market—as Ping-Pong balls.

might by encouraging scientific training and research. Unlike many of their French and British counterparts, German students eagerly put their advanced knowledge to work for the rewards of the market. The organic chemical industry, source of everything from aspirin to TNT, became a German monopoly. Still, Perkin and the British had enjoyed their brief moment of supremacy. It was fitting in 1862 that when Queen Victoria opened a new world's fair in London, though still in mourning for her recently departed Albert, she wore a gown of mauve.

One of the many exhibitors in 1862 was Alexander Parkes. As a metallurgist, craftsman, and inventor, Parkes had been dismayed at rubber's dark, muddy colors. He believed he could do better, and although he did not fully succeed, it can be argued that Parkes paved the way for the great class of materials we know as plastics. In his workshop in the industrial city of Birmingham, England, he began dabbling with nitrocellulose. Made from cotton treated with nitric and sulfuric acids, nitrocellulose could be dissolved in organic solvents such as alcohol and ether and then poured out into a thin film. Frederick Scott Archer had just shown how this solution could be used to coat photographic plates to make glass negatives, but Parkes had other uses in mind: "I determined then to manufacture solid articles principally composed of nitro cellulose and manufacture [them] at the lowest possible cost."

After many years of experimenting with nitrocellulose, Parkes patented a material he called Parkesine. He touted it as good for everything from waterproofing to button making. In that way Parkesine was not unlike rubber, but instead of rubber's blacks, browns, and rusty reds, the new material could be made in bright, shiny colors and even in marvelous imitations of such precious materials as pearl, ivory, and tortoiseshell. At the 1862 London Exhibition, Parkes displayed "PATENT PARKE-SINE of various colours; hard elastic, transparent, opaque, and waterproof." But outside the display halls, Parkesine did not fare so well. Parkes could not decide whether to market it as a cheap substitute for a wide range of materials or as an artistic substance of high quality. Sample goods came back warped and unusable. By the late 1860s Parkes had given up, never able to make his material stable and workable.

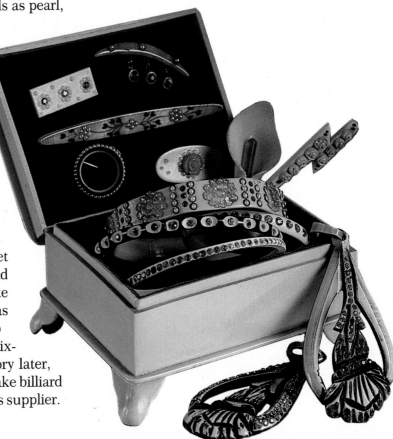

At just this moment, on the other side of the Atlantic, another dabbler stumbled onto the secret that had eluded Parkes. John Wesley Hyatt had worked as a printer in Albany, New York, but like Goodyear and Parkes, he saw himself primarily as an inventor. By 1869 he had set up a company to make small molded items like dominoes from a mixture of wood pulp and shellac. As he told the story later, his real goal was to find a substitute for ivory to make billiard balls—and to win the $10,000 offered by a billiards supplier.

Leo Baekeland

T he aging scientist arrived home with four heavy suitcases. "They were all locked and he had no key," his grandson remembered. "What he had done is lock the first case, put the key in the second, lock the second, put the key in the third, lock the third—and so on. You see, in that way, you did not have *four* bulky keys in your pocket but only *one*. But he had lost that one!"

When the key turned up, the suitcases were found to be full of books and papers, notebooks and manuscripts—Leo Baekeland had forgotten to pack clothes. It was a symptom of advancing age, but also evidence of his lifelong values.

Only toward the end of his days did Leo Baekeland resemble an absentminded professor. Years of work in his chemistry lab in Yonkers, New York, had led him in 1907 to the invention of the first synthetic polymer—the first artificial material made by linking small molecules together to make large ones. Baekeland created a brand new material, a solid like nothing in nature. He mixed the disinfectant carbolic acid (phenol) with the strong-smelling preservative formaldehyde to produce a third substance totally unlike the two originals. It was the first synthetic plastic, and it would change the world.

Marketed under the trade name Bakelite, the stuff could be used a thousand ways. It was molded and tinted into radio cabinets, buttons, billiard balls, subway straphangers, pipestems and cigarette holders, toilet seats, ashtrays, airplane parts. . . .

And so Baekeland started a revolution in science and industry that continues to this day. It sprang from his deliberate search for a profitable solution to the problem of producing synthetic shellac, for Baekeland was both scientist and entrepreneur.

In this iron pressure cooker, Leo Baekeland (upper) made the first commercial batch of synthetic plastic.

He was born in Belgium in 1863 to an illiterate shoe repairman and a maid. His mother, who had observed life outside her cage of poverty, believed that her son's key to escape lay in education. Fortunately, the boy had a brilliant, restless mind, and by the age of 20 he had earned his doctorate maxima cum laude. A professorship gained him entry into the elite world of European academicians. But he had been convinced by reading Benjamin Franklin that there was a place of even greater promise for an educated man of low birth—America.

Before he sailed, he took another step. "My most important discovery at the university was that my senior professor of chemistry had a very attractive daughter. Hence, the usual succession of events." Baekeland's wife, Celine, was the one female he respected too much to call "sillywoman." Yet, when he sent her home to Europe to bear their first child and later delayed her return, he hurt her deeply. Even after their reunion, they often led separate lives.

Meanwhile, Baekeland's research style was taking shape. It combined a willingness to try any tactic with a step-by-step strategy that left little to chance. In the 1890s he produced his first commercial success, a photographic paper called Velox that could be developed in artificial light rather than sunlight. George Eastman of Kodak camera fame was quick to see the paper's potential for amateur photography, and he asked Baekeland to consider selling the rights to it. As the young chemist rode the train to Rochester, he hoped he could get $50,000 for his invention but decided he would take $25,000. Fortunately, Eastman spoke first. He offered $750,000.

With this windfall, Baekeland purchased a Victorian mansion in Yonkers, a suburb of New York City. Here he kept up a lively correspondence with, among others, Henry Ford, the du Ponts, and artist Maxfield Parrish. Nearly every subject intrigued him, from solar energy to wine making.

AT breakfast, your wife pours you a cup of coffee; the handle she takes hold of on the percolator is made of Bakelite, as well as the button under the table she presses for service, and the twin-outlet plug from which are carried the wires to the toaster.

The Material of a Thousand Uses | BAKELITE

A 1920s ad for Bakelite, forerunner of modern plastics.

His grandson recalled walks on which the scientist would tell him the Latin names for plants and animals.

Knowledge was Baekeland's passion. He built a laboratory next to his house and plunged into chemical research. In the early 1900s he took on an elusive problem: finding a way to dissolve the rock-hard substance produced by the reaction between phenol and formaldehyde. Baekeland and other scientists believed that this resin might yield a substitute for shellac, a prized natural substance harvested from insects in Southeast Asia. He knew that a synthetic shellac would be a commercial bonanza. But how to dissolve the brittle resin on the bottoms of his laboratory beakers?

Then one day Baekeland turned his thinking inside out: Could he make use of this frustrating resin? Eventually, by controlling the conditions of the reaction, he transformed it into Bakelite, a hard, clear solid that was impervious to acids, electricity, and heat and that could be dyed bright colors. With characteristic good timing, he was able to supply what the new electrical and automotive industries needed—insulators and heat shields. Manufacturers of everything from knife handles to telephones had no trouble finding a myriad other uses for this wonderful, versatile substance.

Bakelite's creator controlled his industry, becoming a multimillionaire and spending more and more time at his home in Coconut Grove, Florida, where he retired in 1939. Known as something of a miser, he chose now to live the simple life he preferred, sleeping in a sparely furnished room and eating his meals out of tin cans. He grew increasingly untrusting, intolerant, and withdrawn, shunning even the company of his family.

He took to dressing all in white, like Mark Twain. When the Florida sun made him too hot, he would simply walk into the swimming pool with all his clothes on—white sneakers, white shirt and duck trousers, and sun hat. "The evaporation keeps you cool," he would exclaim. To the end, Leo Baekeland was the practical scientist.

Jerry Camarillo Dunn, Jr.

Bakelite objects: a gear, razor holder, salesman's samples, deodorant jar, salt and pepper shakers, and cigarette box.

Like others, Hyatt turned his attention to nitrocellulose. He discovered that shredded nitrocellulose could be mixed with camphor and heated under pressure to produce a tough whitish mass that kept its shape. It could be drilled, cut, sawed, and, when warmed, bent and molded. After a few more experiments, Hyatt's brother Isaiah dubbed the new material celluloid, and the two proceeded to seek out markets.

Celluloid succeeded where Parkesine had failed, largely because it was a better product, but also because the pragmatic Hyatt brothers had a better grasp of what their material could—and could not—do. Some of this they learned the hard way. Celluloid dental plates were an embarrassing failure. Unhappy users complained of the ever present taste of camphor or, worse, of plates that warped after a gulp of hot tea. Legend has it that when celluloid billiard balls bumped into each other during play, they sometimes exploded with a gunshot-like sound, causing hustlers in one Colorado saloon to draw their guns.

Celluloid's most important quality was the ease with which it could be made to look like other materials. Dyed, it resembled marble or coral. It took the effect of mother-of-pearl wonderfully, and swirling in browns and oranges produced imitation tortoiseshell. Soft celluloid pressed against linen took on the texture of a weave—perfect for collars and cuffs. The best effect was that of ivory, for clever techniques could reproduce in celluloid the striations so distinctive in an elephant's tusk. Since celluloid was not as hard or dense as ivory, it never made good billiard balls. But for just about everything else, from piano keys to combs, celluloid's ivory looked like the real thing. Consumers came to perceive the first successful plastic as a cheap, practical imitation of other finer, more natural substances—an identification that lingers even with the more sophisticated plastics of our day.

Celluloid stands alone as the only important man-made plastic of the 19th century. The true age of plastics emerged just after 1900, the creation not of craftsmen like Parkes and Hyatt, but of chemists, the heirs of Liebig and Hofmann. When Leo Baekeland announced to the world his invention of Bakelite in 1909, he did so in the proper professional setting of the American Chemical Society. Perhaps as an ironic tribute

May 15, 1940: Nylon stockings premier in New York City; four million pairs are snapped up in hours. Two years later nylon, the world's first man-made fiber, went to war in parachutes and tire cords—and enough tent fabric to cover Manhattan.

The man who started it all was Wallace Hume Carothers (right), a Du Pont chemist who linked small molecules into long chains, or polymers, and transformed them into fibers stronger, cheaper, and more elastic than silk. His breakthrough—nylon—spawned an endlessly growing family of synthetic fibers: Ban-Lon, Lycra, Fortrel, Antron, Orlon, Dacron. . . .

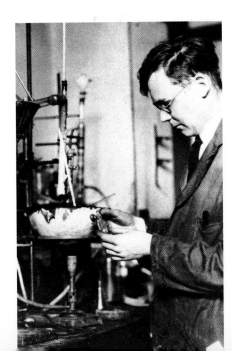

to celluloid and its beginnings, Baekeland pointed out that his material "makes excellent billiard balls." And, indeed, the synthetic resin that Baekeland had first concocted in his backyard laboratory in Yonkers, New York, two years earlier proved an excellent substitute for billiard ball ivory, as well as for an astonishing range of other things.

Bakelite was the first phenolic resin, a material whose resistance to heat, electricity, and chemical action made it perfect for insulating and protecting a host of products, from cookware to telephones. As a binder for materials like sawdust, fabric, or paper, phenolic resins opened the way for new composite materials, from utilitarian plywood to colorful Formica laminates. Bakelite's moldability and durability made it popular for bracelets and beads and radio cabinets.

Bakelite's real importance lay not so much in what it did but in what it was. Leo Baekeland had created the first true synthetic plastic when he mixed the simple organic chemicals phenol and formaldehyde in his big iron reaction vessel. In this vessel (he called it a Bakelizer) he was able to control the chemical combination to produce what we would call a synthetic polymer. In fact, it took another decade for chemists to agree on the existence of the giant molecules known as polymers, but

Baekeland showed how the incomplete theories of his day could be combined with empirical knowledge to produce new plastics.

The 1920s and '30s saw an outpouring of plastics as large corporations put their faith and their money into a burgeoning market. In 1926 a chemist for rubber manufacturer B. F. Goodrich discovered how to make adhesives and sheets from polyvinyl chloride, and vinyl became one of the most common plastics. Four years later the giant German firm I. G. Farben put polystyrene on the market, providing a hard, shiny material that could be molded into refrigerator liners, tackle boxes, and a host of other items. Another German company, Röhm and Haas, brought out Plexiglas in 1935 (a British firm developed Perspex at about the same time), and soon the crystal-clear plastic found its way into furniture, jewelry, clocks, aircraft windows, car taillights, boat windshields, and camera lenses. New materials changed styles and fashions in the 1930s, and streamlining influenced the shape of everything from vacuum cleaners to locomotives.

By the 1930s chemical theory was catching up to industrial practice, and some chemists became even bolder in their attempts to mimic and outdo nature's own materials. The most famous product of this boldness was nylon. In 1928 the giant Du Pont Company, already in the business of manufacturing plastics and fibers, recruited a bright 32-year-old chemist, Wallace Hume Carothers. Du Pont's promise to provide Carothers with the best equipment and assistants persuaded him to abandon his teaching post at Harvard. Du Pont assured him that

Like a flying saucer, the world's first plastic house beckoned postwar Americans into realms strange and new at Disneyland. In the House of the Future's kitchen, a polarized plastic ceiling glowed overhead, and shelves rose out of counters at a finger's touch. Ten years and 20 million visitors after construction in 1957, the house's cantilevered fiberglass wings sagged only a fraction of an inch. The House of

the Future came down in 1967, when the present caught up with the future.

Pastel plastic drinking tumblers and storage containers first appeared at the end of World War II, the brainchild of chemist Earl S. Tupper. Sold in living rooms across America, Tupperware Lettuce Crispers and Saltine Savers became household staples, the home parties an American institution that grew into a worldwide empire.

he would be free to define his research tasks, asking only that he investigate polymerization—building giant molecules out of smaller ones.

Carothers helped Du Pont scientists produce a synthetic rubber called neoprene, then turned his attention to making synthetic fibers through polymerization. Scientists had made new fibers, like rayon, from cellulose but had never synthesized them from simple chemicals. After trying different chemicals, Carothers and his assistants finally settled on amides—the same chemicals that form the protein in wool, silk, and other animal fibers. By carefully controlling the reaction, the scientists built up polyamides into a strong, flexible fiber.

In mid-1938 yarn spun from a polyamide was knit into experimental stockings, and the following year nylon went into production. The public reacted with unprecedented enthusiasm as word went out that nylons would not run or tear and lasted forever. The truth dampened enthusiasm a bit. Consumers met with further frustration when, in December 1941, all nylon production was diverted to wartime needs. Unfortunately, Wallace Carothers did not live to see nylon's success, for in April 1937 the often moody chemist, who always carried a small vial of cyanide, took his own life.

The Second World War shaped the future of the plastics industry. Wartime shortages of plastic forced manufacturers to seek substitute materials, but after the war they began cranking out an overwhelming variety and quantity of plastic goods. Eager to stake out new markets, many companies misjudged both their materials and their customers, often making too much that was too cheap. Although consumers became disillusioned with the vision of a world remade with artificial materials, chemists continued producing their miracles at an ever faster pace. Spurred on by the needs of electronics, defense, and, eventually, space technologies, researchers created materials with ever newer, ever stranger properties, from a substance that seemingly defied friction (Teflon) to one that stopped bullets (Kevlar). A century after the Crystal Palace, the very feel of the world was different.

New metals also helped reshape the world. Aluminum, now a common and inexpensive metal, was just a fascinating curiosity in the mid-1800s. Metallic aluminum does not occur in nature, and scientists did not even suspect its existence until after 1800. Some of the best chemists labored to produce tiny samples. Hans Ørsted, famous for discovering electromagnetism, eked out a small lump of aluminum in 1825, and 20 years of experiments by the great German chemist Friedrich

The 1987 America's Cup winner Stars and Stripes *heels to port under sails of Kevlar, a tough, lightweight synthetic fiber that resists stretching. Five times stronger than steel and three times stiffer than fiberglass, Kevlar was discovered in 1965 by Stephanie Kwolek, a Du Pont chemist searching for a fiber stronger and stiffer than nylon.*

Du Pont hailed Kevlar as the most important man-made fiber since nylon. *Made into lightweight vests, Kevlar stops bullets. Woven into cables, it anchors supertankers and offshore oil rigs. Sandwiched into the fuselage of the aircraft* Voyager, *it flew nonstop around the world in 1986.*

Once as valuable as silver, aluminum made its debut in 1855 at the Paris Exposition. A year later the costly curiosity, adorned with gold, was cast into a baroque-style centerpiece (opposite) and presented to Napoléon III. European scientists had extracted small amounts of aluminum from ore in the early 19th century—but at great expense. A breakthrough came in 1886, in an Ohio woodshed and a French tannery. Two months apart, two 22-year-old inventors—Charles Martin Hall and Paul Héroult—discovered the economical method still used today to smelt aluminum.

In 1888 Hall (shown standing at far right, with co-workers and their families) helped found the company that became Alcoa—Aluminum Company of America. Alcoa safeguards Hall's first pellets of pure aluminum (lower), dubbed the "crown jewels."

Wöhler yielded some impure aluminum globules. In the 1850s one of the century's most prolific chemists, Henri-Étienne Sainte-Claire Deville, was treating aluminum chloride with potassium when he noticed tiny nuggets of metal. Sure enough, they turned out to be almost pure aluminum, and Deville instantly realized their importance.

When he announced his results to the august French Academy of Sciences in early 1854, Deville declared his intention to commercialize his discovery. Captivated by the silvery yet incredibly light metal, the academy voted a grant of 2,000 francs to help him. Trying electrical methods (which turned out to be too expensive because they depended on batteries) as well as new chemical approaches (substituting cheaper sodium for potassium), Deville produced several large bars of pure aluminum in time for display at the 1855 exposition in Paris. Alongside the bars, billed as "silver from clay," lay aluminum jewelry, trinkets, and a baby rattle for Napoléon III's son. The emperor, dreaming of an army made light and swift by the featherweight metal, urged Deville on.

A small aluminum industry struggled into existence in France. Though still interested in promoting the material, Deville returned to his college laboratories, concluding that the problems of commercialization were not the proper concern of an academic man. The more practical sorts who took up the work eventually improved Deville's processes and reduced the price of aluminum from 1,000 francs per kilogram to 130 francs by 1872. That year the French produced almost two tons and an English manufacturer about three-quarters of a ton. For the next decade such figures remained typical. After 30 years of work, the world's annual aluminum production amounted to only two and a half tons.

More aluminum could have been made, but it simply wasn't wanted. The continuing high price certainly dampened demand, but it didn't explain everything. The "silver from clay," touted by technical writers in Europe and America as the metal of the future, actually had few practical uses in the 19th century. Aluminum didn't tarnish like silver, so it made better egg cookers and mustard spoons. Its light weight recommended it for sextants and for other

From the panels of New York's World Trade Center (opposite) to bottle caps, gum wrappers, and gutters, aluminum is everywhere. The most plentiful metallic element in the earth's crust, it is extracted from ore, then cast, rolled, forged, or drawn into thousands of shapes. Extruded aluminum (above), squeezed through dies like dough through a pasta maker, can turn into beer cans or rocket fuel tanks.

But when aluminum first entered the marketplace in the 1880s, no one knew quite what to do with the new metal. Aluminum pots and pans cost six times as much as tin cookware. Long-lasting, lightweight aluminum shoes? The idea didn't catch on.

Then, in 1903, an engine with an aluminum crankcase powered the Wright brothers' first plane aloft. The aluminum industry took off too, and today the world uses more aluminum than any metal except iron and steel.

instruments, as well as for fine opera glasses or even fancy fans. Aluminum jewelry enjoyed some popularity, and special uses, such as the 100-ounce apex of the Washington Monument (installed in 1884 after being displayed at Tiffany's), brought occasional celebrity.

As a material of great promise, aluminum was an obvious target for the would-be inventor, the ambitious young man or woman seeking wealth and reputation through an imaginative technical feat. In 1886, 22-year-old Charles Martin Hall had just graduated from the local college in Oberlin, Ohio. That same year, Paul-Louis-Toussaint Héroult, just a few months older and a former student at the School of Mines in Paris, was working in his father's tannery in Gentilly, a Paris suburb. Both young men had studied with chemistry professors who turned them to thoughts of the wonderful things that could be done with materials by discovering the right chemical tricks. And both were fascinated by aluminum. Early in 1886 these two, independently, solved the key technical problems that stood in the way of cheaper aluminum.

The Hall-Héroult process used electricity to extract aluminum from its oxide. While experimenters had tried this for years but failed, the new process succeeded for several reasons. By the mid-1880s the dynamo could produce large quantities of cheap electricity. Even more important, the new process used cryolite, an easily melted mineral, to dissolve the aluminum oxide, and carbon to line the reaction crucible. These materials allowed the electric current to separate aluminum from oxygen without the interference or contamination encountered

What makes stainless steel stainless? Why is it the material of choice in a diner's kitchen? A thin, transparent film of iron oxide and chromium prevents soap, water, food, and air from eating away the metal underneath. Its smooth, hard surface doesn't trap dirt, bacteria, or molds, making it hygienic. And stainless steel heals itself. If a toaster or coffee maker is nicked or scratched, the protective oxide forms a new shield almost instantly.

A search for better cannon linings led to the discovery of chromium's ability to generate an oxide coating. In 1913 British metallurgist Harry Brearley noted that steel made of iron and chromium withstood attack by different chemical solutions. Knives formed from this steel did not rust, and thus was born the stainless steel industry. By 1947 American factories were producing 14,000 tons of stainless pots and pans a year (above).

by other researchers. Pure metal settled to the bottom of the crucible.

By mid-1888, factories using the Hall-Héroult process were in full swing. A Swiss company began operating under Héroult's patents later that year, and the English, French, and Germans soon followed. Charles Hall found a group of investors in Pittsburgh to back an enterprise that set itself up as the Pittsburgh Reduction Company. With just a couple of assistants, Hall managed to turn out about 50 pounds of aluminum a day. He started with a selling price of five dollars a pound—almost one-third the cost of the cheapest aluminum then available. It wasn't cheap enough. Bars of the metal started piling up in the company offices. Hall reduced prices as low as two dollars a pound for anyone willing to purchase half a ton. Still, the company's best customers remained the Pittsburgh steelmakers, who discovered that a few pounds of aluminum thrown into several tons of molten steel absorbed harmful gases. It would take more than a new process to create the aluminum age touted by some scientists and journalists.

As with most new materials, the popularization of aluminum required imagination, chutzpah, and foot-in-the-door salesmanship. When a cookware maker agreed with Hall that the new metal made fine tea kettles but protested that he couldn't work with it, the Pittsburgh manufacturers found themselves in the cookware business in the early 1890s. Gradually others were persuaded to experiment. The military tried aluminum for everything from mess kits (very good) to horseshoes (terrible). Physicians made artificial limbs, manufacturers produced watch hands and wagon frames, and hairdressers tried aluminum combs and bobby pins. All this didn't amount to much.

The future of the aluminum industry lay instead in the new technologies of the 20th century. First, the infant electrical industry took advantage of aluminum's virtues as a conductor for transmission wires. Then the lightweight metal went airborne with the Wright brothers in 1903—and has stayed there ever since. By the 1930s aluminum could be found almost everywhere, from pots and pans in the kitchen to the trim on skyscrapers and streamlined locomotives.

The evening of May 26, 1934, a streamlined train pulled up to a platform at Chicago's lakefront, creating a sensation among the crowd gathered for the opening of the second year of the Century of Progress Exposition. The Chicago, Burlington and Quincy Railroad's *Zephyr* had left Denver just after five o'clock (Chicago time) that morning and sped to Chicago in 13 hours, half the time of other trains, hitting a top speed of 112 miles an hour. Not only the *Zephyr*'s speed record caused a sensation. Gleaming stainless steel accentuated its radical streamlined shape, embodying society's growing fascination with dazzle and glamour in buildings, machines, and even everyday objects. The new train featured other novel materials—spun aluminum in the seats and room dividers, Formica tabletops, chromium fittings—but Burlington publicity heralded the *Zephyr* as "A Symphony in Stainless Steel."

More than any other material, steel had laid the foundation of industrial technology. Remarkable innovations in steelmaking in the second half of the 19th century accelerated the spread of industrialization. And as steel displaced wood, stone, and older forms of iron in everything from buildings to ships, scientists sought ways to improve the material. In particular, making a rustless or stainless steel was an old dream, and chemists and metallurgists tried alloys of every variety. One of the most promising alloying materials was chromium, a metal discovered only in the late 18th century. In the last decades of the 19th century, scientists observed that small amounts of chromium toughened steel, but the right formula for corrosion resistance proved elusive. Finally, in the years just before World War I, investigators in both Europe and America discovered several types of chromium steel that resisted corrosion, staining, or any kind of chemical action. It took another decade of research to determine how best to use these stainless steels, but by the 1930s architects and designers had at their disposal a remarkable substance, combining the strength and sturdiness of steel with the imperviousness of aluminum and the gleam of silver. Even more important, perhaps, they had available a highly visible symbol of the new materials that would remake the look of the 20th century.

In the half century since the *Zephyr* made its dramatic appearance on Chicago's lakefront, the number and variety of new materials have exploded. Not only have new plastics and alloys continued to make places for themselves in the world, but even materials of great antiquity have been changed into novel substances. Glasses that will withstand rapid temperature changes, ceramics sturdy enough to protect a spacecraft from the searing heat of atmospheric reentry, crystals that allow electrons to be controlled with such fineness that computer processors are reduced to thumbnail size—all are results of knowledge and imagination combined in ways that are remaking the very stuff of our world. Sometimes the remaking is not all to our liking, as when we surround ourselves with wastes and by-products that threaten to injure or kill either us or our fellow species, or when we find ourselves yearning for a more natural environment. The remaking will not stop, however. Our world and that of our children will look, feel, and behave differently from that of our ancestors, and living with it may require as much imagination as creating it.

Harder than steel, lighter than aluminum, stronger than glass, Pyroceram withstands the shock of ice and flame in cook-serve-freeze Corning Ware (far right) or on the nose cone of a guided missile. Corning Glass Works heralded its creation of Pyroceram in 1957, claiming that this nonporous ceramic formed from glass was one of the greatest advances in glass research since Corning introduced Pyrex bakeware (right) in 1915. Derived from the Greek word pyr *for "fire," Pyrex was originally developed as a heat-resistant glass for railway lanterns, which often cracked when taken out in the rain or snow for signaling.*

Capturing the Image

By Thomas B. Allen

magine a battle in the Revolutionary War. Now envision the Civil War...World War I ...World War II...Vietnam. For the Revolutionary War, your mind had to summon up sketches, drawings, paintings. For the other wars, your mind's eye produced a different kind of image: photographs. The camera, unlike any other invention of modern times, altered perception. And, as the camera changed, so did the perception of reality. The grainy, black-and-white tableaux of the Civil War and World War I evolved into the talking, fast-moving, blood red narratives of modern war. Images, once sketched and silent, could cheer and scream, march and fall.

Cameras have put newsreels of war in our minds, along with the softly focused scenes of peace. Through our minds parade images of events both historical and personal: the inauguration of a President, the wedding of a friend, the step of a man on the moon, the first steps of a child.

From cave walls to Victorian parlor walls, the only visions of reality on display were the works of human eye and hand. Then, in a burst of inventive energy, photography interposed its apparatus and its chemical sorcery between the hand and eye. At a time when people were beginning to realize that machines would change their lives, the Victorian era produced the modern camera. Its images would force people to change profoundly the way they saw the world and themselves.

When the camera first appeared, many illusions of reality disappeared. But in its evolution the camera would one day become the master of illusions, portraying make-believe worlds on a new kind of cave wall. Photography, like light itself, moved with astonishing speed and dazzled its awed beholders. What began as a novelty, a toy that froze time, became a force that blurred reality.

Observations of Aristotle led to the realization in ancient times that sunlight streaming through a small hole in the wall of a darkened room would project an inverted image of the sun onto the opposite wall. The phenomenon remained an intermittent curiosity until about the 16th century, when artists and draftsmen began putting a lens in the hole on the wall of what by then was known as the camera obscura, the dark room. By moving a paper backward and forward between the lens and the opposite wall, artists in the dark room could focus images from the sunny outside world. "There on the paper," a Padua academician wrote in 1568, "you will see the whole view as it really is, with its distances, its colors and shadows and motion, the clouds, the water twinkling, the birds flying." An artist could trace the image with a pen and produce a vivid image only once removed from reality.

By the next century the dark room had shrunk to a dark box an artist could carry around. Inside it, improved lenses and mirrors produced

sharp, upright images that a steady hand could trace, whether it was a wide landscape to be re-created in miniature or a person sitting for a life-size portrait. Variations on the camera obscura gave artists a tool for their hands. With this tool the hand could seize a version of reality. But the reality had no depth. An artist who took a magnifying glass to the sun-given image saw details too dense or too tiny for the human eye to see or most human hands to sketch—a leaf's network of veins, a blossom's sprinkle of pollen. If only there were a way to grasp those exquisite images of reality, a way to make sunlight the artist.

There was a way. In the early 1700s a German scientist discovered that the rays of the sun darkened the salts of silver; the more intense the sunlight, the darker the salts. Discoveries about light-sensitive chemicals continued. Just before the century ended, Thomas Wedgwood, youngest son of the great British potter Josiah Wedgwood, began experimenting with silver nitrate.

Thomas knew about the camera obscura, used by the Wedgwood firm to draw sketches on china. He hoped to project images onto sensitized surfaces and eliminate the need for tracing. He placed an object on a paper moistened with silver nitrate and then exposed the paper to sunlight. When the object was removed, its silhouetted image remained. But, because the entire paper remained sensitive to light, the images he produced darkened when exposed to light until the entire paper turned black. He never found a way to hold the image.

Within a generation, Wedgwood's experiments, like his images, had faded away. But another British experimenter took up the quest. William Henry Fox Talbot, a brilliant scientist and amateur artist, had often used a camera obscura to aid his sketching. In 1833, while thinking about the startlingly beautiful "fairy pictures" projected onto his drawing paper, he decided to find a way to "cause these natural images to imprint themselves durably, and remain fixed upon the paper."

After several experiments, Talbot mixed a weak solution of table salt (sodium chloride) and water, soaked sheets of paper in the solution, and dried the paper. He then daubed silver nitrate on the paper. The chemicals formed silver chloride, which saturated the paper, making it sensitive to light. On the paper Talbot placed translucent objects, such as feathers, leaves, and lace. The objects blocked the light that blackened the rest of the paper and left behind delicate white silhouettes— negatives of reality. He preserved these images by soaking the paper in a concentrated solution of table salt.

"Lights! Camera! Action!"—and, a movie director might well add, "Inventions!" From camera obscura to Technicolor, photography has marched to a steady beat of inventions. Born as a way to preserve reality, photography evolved into an art that, in the movies, produced illusion. Here, on a 1950s set for The Seven Year Itch, *powerful lamps play a role as old as photography: bathing an image in light.*

Talbot kept experimenting, trying various chemical coatings. One day in September 1840, while testing the sensitivity of several papers, he exposed them in a camera obscura, by then simply called a camera. Sometime later, examining a paper by candlelight, he was "very much surprised to see upon it a distinct picture." Earlier, he had found that by placing the translucent paper on another sensitized sheet and exposing the union to light, the original negative image was transferred, as a positive image, to the second paper.

Talbot's discovery led him to the process that is still the basic principle of photography: Latent images are produced by exposing sensitized coatings to light in the camera, and then these negatives are used to make positive prints. Drawing on his classical knowledge, he dubbed his process calotype after the Greek word for "beautiful" and patented it in 1841. But, unknown to Talbot, long before he got his patent, Joseph-

The world's first photograph from nature, a ghostly image of light and shadows shows the French estate of Joseph-Nicéphore Niépce. About 1827 he aimed a lens-in-a-box camera out an upper window of his home and for hours let sunlight flow onto a pewter plate. He had thinly coated it with etchers' bitumen of Judea, an asphalt that hardened when exposed to light. After long exposure, Niépce rinsed the plate in a solvent. The image showed what his "kind of artificial eye" had seen: a courtyard, a barn's slanting roof, a pear tree. Undissolved asphalt shows as light tones. The dark tones are those shiny surfaces of the plate where the unexposed asphalt had been dissolved by certain oils.

Nicéphore Niépce had made the world's first photograph.

The French brothers Joseph and Claude Niépce had been tinkering with things mechanical since childhood. As adults, their most spectacular creation was a boat noisily propelled by an engine fueled with explosives. While Claude worked on promoting their pyrotechnic engine, Joseph took up the quieter work of advancing an invention he named heliography, drawing with the sun.

About 1814 Joseph found a way to produce images on pewter plates covered with light-sensitive varnishes that he made himself. Sometime around 1827 he aimed his camera out a window at a sunny courtyard and for about eight hours let light pour through the lens. The image Niépce developed was dim, cloudy—and historic. But his photographic milestone stood on a dead-end road, heliography. The exposure time was too long, the image too faint.

Asleep on a Paris rooftop with his camera nearby, a daguerreotypist is lampooned for the quirks of his new hobby: long exposures and a tendency to focus on such motionless subjects as roofs and chimney pots. Louis-Jacques-Mandé Daguerre (below, in an 1844 daguerreotype) invented his namesake process. The camera he promoted (lower) started a photography craze that would click throughout the world.

Niépce soon realized that if his work were to go on, he would have to collaborate with another Frenchman trying to capture images, an impetuous, resourceful painter by the name of Louis-Jacques-Mandé Daguerre.

"I am afraid he is out of his mind," Daguerre's wife said of her husband, despairing of his days and nights without sleep, his obsession with the "idea that he can fix the images of the camera." Daguerre's idea, however, looked promising to Niépce. After some cautious approaches, Niépce met with Daguerre and in 1829 signed a ten-year partnership agreement. Four years later Niépce died, virtually penniless. He had made the world's first photograph. But it would be Daguerre who would make photography practical and popular. His very name would, for a time, become the world's word for photograph.

William Henry Fox Talbot

Neighbors watched suspiciously as bottles of chemicals and stacks of paper arrived at William Henry Fox Talbot's establishment in Reading, England. Someone must be forging money or concocting something nefarious. The peepers had guessed wrong: Not counterfeiting but "magic realized—natural magic" was Talbot's business in the mid-1840s. Soon workers were printing photographs and illustrating a commercial book with them for the first time in history.

A decade earlier, Henry Talbot had discovered the principle on which modern photography depends: Negatives can print positives. In the 1850s he invented the basis for photogravure, eventually enabling photographs to be reproduced by the millions. A brilliant scholar whose inheritance supported his research, Talbot made major contributions in mathematics, botany, astronomy, optics, and chemistry. He designed windmills and electric motors. He was a leading translator of Assyrian cuneiform, wrote books on folklore and etymology, and served in Parliament. He was a restless intellectual voyager who aimed to penetrate the "ultimate nature" of whatever he was studying, hoping thereby to bring humanity a step closer to understanding the universe. Talbot saw no need to specialize, since he believed that all knowledge is interrelated.

Born in 1800, Henry Talbot would have started life at Lacock Abbey, Wiltshire—owned by his forebears since 1539—had hard times not befallen the family. Henry's father, an officer in the dragoons, died only months after Henry's birth, leaving behind a trail of debts to "appal the stoutest heart," according to a document of the time. Henry's mother, Lady Elizabeth Fox-Strangeways, salvaged her son's inher-itance by renting Lacock Abbey, raising Henry in relatives' houses, and marrying a prosperous man.

At age six the future photographic pioneer first demonstrated his compulsion to chronicle his accomplishments. Sent to boarding school two years later, he requested his stepfather to "Tell Mamma and everyone I write to to keep my letters and not burn them."

Lady Elizabeth, who complied, worked throughout her life to ensure that nothing kept him from the great achievements she saw as his destiny.

By the time Henry entered Harrow in 1811, he was writing to his mother in French punctuated with Greek and Latin dictums. He predicted eclipses and became enchanted by chemistry. One of his experiments, conducted at his housemaster's home, "exploded with the noise of a pistol and attacked the olfactory nerves of the whole Household," Lady Elizabeth recorded.

"He has an innate love of knowledge and rushes towards it as an otter does to a pond," reads an evaluation of Henry Talbot written shortly after he graduated from Cambridge in 1821 with a prize in Greek and first-class honors in mathematics. But for all his intellect, Talbot could not draw. On his Italian honeymoon in 1833, Talbot tried to sketch the scenery at Lake Como. Discouraged, he reflected on the "inimitable beauty of . . . nature's painting" displayed in a camera obscura and resolved to experiment with silver nitrate on his return to England. For his first "photogenic drawings" Talbot pressed leaves and lace against salted paper and exposed them to sunlight. Soon he paid a local carpenter to build small wooden cameras and produced "very perfect but extremely small pictures; such as . . . might be . . . the work of some Lilliputian artist."

"Mousetraps," his wife called the cameras that Henry scattered around Lacock Abbey and its grounds. Talbot realized that his photogenic drawings, reversed with respect to light and dark, right and left, could be used as negatives to print naturalistic images.

To Talbot, photography meant the union of science and art: With a camera, the scientist could become an artist; the artist would enlist chemistry and optics. Photography would reveal "a multitude of minute details which add to the truth and reality of the representation, but which no artist would take the trouble to faithfully copy from nature." In Talbot's own pictures he constantly varied lighting and angles in a quest to discover the "ultimate nature" of his invention.

Unfortunately, Talbot procrastinated in disclosing his discoveries and was stunned by the announcement of Louis Daguerre's work in 1839. Faced with "the loss of all my labours," Talbot sought to establish himself as the prior inventor of photography by having his pictures shown at the Royal Institution in London. But the public, to Talbot's frustration, still gave Daguerre credit

In a panorama of Talbot's studio in Reading, England, cameramen photograph artwork, a seated subject, and statuary. The aproned man prints images by sunlight.

Talbot made this calotype of his mother in 1842.

for the invention of photography.

Gradually, Talbot's focus shifted to commercialization. In 1841 he patented the calotype, a process that slashed exposure times to minutes, or sometimes seconds, by enabling the photographer to chemically develop a latent image. Talbot craved recognition more than money, but he charged a license fee to calotype photographers, and his eagerness to prosecute those he considered patent offenders all but stifled the development of paper photography in England. After an emotional legal wrangle, Talbot decided not to renew the calotype patent in 1855.

In later years Talbot continued his research in mathematics, spectroscopy, and Assyriology. Today his quest "to make space and light . . . speak to us" lives on in all photography.

Carole Douglis

156

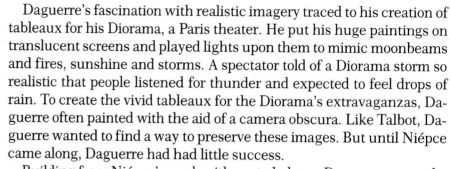

Daguerre's fascination with realistic imagery traced to his creation of tableaux for his Diorama, a Paris theater. He put his huge paintings on translucent screens and played lights upon them to mimic moonbeams and fires, sunshine and storms. A spectator told of a Diorama storm so realistic that people listened for thunder and expected to feel drops of rain. To create the vivid tableaux for the Diorama's extravaganzas, Daguerre often painted with the aid of a camera obscura. Like Talbot, Daguerre wanted to find a way to preserve these images. But until Niépce came along, Daguerre had had little success.

Building from Niépce's work with coated plates, Daguerre covered a plate of copper with highly polished silver. Then he let iodine vapors play over the plate in a closed box for as long as half an hour. Next, he put the plate, still shielded from light, into a lens-equipped camera obscura. He aimed it at an inanimate subject, such as a statue, and removed the lens cover. The light that entered the camera fell upon the silver iodide on the plate. The more light bathing the subject, the greater the reduction of the silver iodide to silver. This chemical reaction formed a latent, or invisible, image.

Exposure demanded a great deal of light—"five to six minutes in summer and from ten to twelve minutes in winter," Daguerre said. After what he judged to be sufficient exposure, he removed the plate from

the camera and developed it by putting it in a box containing heated mercury. The vapors joined with the reduced silver to form an amalgam, thus revealing the image. But the image could not be exposed to light because light-sensitive chemicals still remained on the plate.

Luckily, Daguerre knew about the discovery made by Sir John Herschel, a British scientist and astronomer. Herschel had found that sodium thiosulfate, then known as hyposulfite of soda, would fix images. Photographers still call the fixer hypo.

Daguerre bathed the plate in hypo. He then washed the plate in distilled water, dried it, and put it under glass to protect the thin film of mercury amalgam. He named this preserved image the daguerreotype. Its image was reversed, a mirror. But a mirror like none before, a mirror whose fleeting image was frozen in time.

Viewed from some angles, the image disappeared, making the shimmering daguerreotype faces and figures seem ghostly, unfathomable. A German writer called the images "blasphemy . . . an invention of the Devil." But Edgar Allen Poe, looking at those entrapments of time and memory, saw "a more absolute truth, more perfect identity" in the image than in the reality it portrayed.

France acquired the invention at a cost of an annual pension of 6,000 francs for Daguerre and 4,000 francs for Isidore Niépce, son of Joseph.

Images at photography's dawn: An affectionate trio (opposite, upper) animates a multiple carte de visite, *portraits sometimes used as calling cards. These were made by a multi-lens camera (opposite, lower) that exposed a series of poses on one plate and speeded up production. The stereoscopic viewer added dimension to family albums, which began small, with oval-bordered tintypes, then swelled in size and Victorian opulence. Jewelry also enframed loved ones immortalized on tintypes and collodion prints.*

On August 19, 1839, the secrets of the process were revealed at a joint meeting of the French academies of science and art, foreshadowing the still debated question: Is photography art or technology?

Some artists—Toulouse-Lautrec, Degas, and Delacroix among them—became interested in photography and did not see it as competition to painting. But a group of painters in France signed a manifesto that said photography could not "in any circumstance" be considered a work of the intelligence or an act of art.

Samuel F. B. Morse was one portrait painter who did accept the new mirror. Soon after Daguerre's process became publicly known, Morse happened to be in Paris patenting the telegraph he had invented. He called on Daguerre, saw images he called "Rembrandt perfected," and wrote a fervent description that his brothers, New York newspaper editors, published. Morse's report about Daguerre's discovery was reprinted in newspapers throughout the United States.

Back home, Morse, using a camera made by the man who produced his telegraphic instruments, learned photography. Morse set up a glass-roofed photographic portrait studio, which helped to sustain him while he awaited government support for his telegraph. He also taught

Roger Fenton, pioneer war photographer, poses on his horse-drawn darkroom, needed because wet plates had to be developed immediately. In 1855, during a Crimean War battle, a cannonball holed the van, letting in light and ruining plates Fenton was processing. Early photographers inspired a lasting taste for realism. Photographs crowding the walls of a British exhibition (above) in 1884 were acclaimed as a new form of art. Most were family portraits and landscapes.

photography. One of his first students was Mathew B. Brady, whose cameras would starkly immortalize the battlefields of the Civil War.

The American demand for daguerreotype portraits created an industry overnight. Around the time Morse established his New York studio, daguerreotypists went into business all over the country and soon were turning out about three million images a year. The photo portrait craze created a need for many inventions—from special portrait lenses to diabolical head-vises that kept people still during long exposures.

By about 1860 the daguerreotype was doomed by what would become a lasting photographic process: a transparent negative from which prints could easily be made. One discovery involved guncotton, a flammable substance made by drenching raw cotton with sulfuric and nitric acids. Dissolved in ether and alcohol, guncotton produced a sticky liquid that dried into a tough, transparent film. It was called collodion, from the Greek word for "adhere."

In 1851 Frederick Scott Archer, a British sculptor who photographed his models, began the march toward faster, easier photography. Archer added potassium iodide to collodion, coated the mixture on a glass plate, bathed it in silver nitrate, and exposed it in a camera. The plate

George Eastman

When George Eastman was a boy, he converted wire knitting needles into an intricate puzzle. A friend, fascinated by the creation, asked if he could have it. George offered to sell it for a dime. The friend became furious, but George was adamant. He held out for—and got—his ten cents.

From childhood, George Eastman displayed traits of the quintessential entrepreneurial inventor—a mind capable of complex ideas, the manual skills of a tinkerer, business acumen, the courage of his convictions, and a strong appreciation for money. He was born in Waterville, New York, on July 12, 1854, to colonial stock. His father owned a nursery and commuted 150 miles westward to Rochester, where he had established Eastman's Commercial College. George's mother, Maria Kilbourn Eastman, was an educated woman whose favorite aphorism George often quoted: "Talk is cheap, but it takes money to buy whiskey."

In 1860 the Eastman family moved to Rochester. Two years later George's father died suddenly. The family lived on in the fashionable neighborhood, and George began his studies at Mr. Carpenter's, "the best private school for boys in the city." To make ends meet, Maria was forced to take in boarders. George adored his mother, and seeing her mop floors and make other people's beds stirred in him a magnificent obsession: "All I had in mind," he said in later years, "was to make enough money so that my mother would never have to work again."

He left Mr. Carpenter's in 1868 at the age of 13 for a job as office boy in an insurance agency at $3 a week. He swept the floor, cleaned cuspidors, and meticulously recorded in a notebook every penny he earned and spent, including 65 cents squandered for ice

Tourist George Eastman aims his Kodak in 1890.

cream on his 14th birthday. By 1876 he was earning $1,400 a year as a bookkeeper. His notebooks grew fat with pages of entries: coal and gas for the house, flute and dancing lessons, hats, cigarettes, candy, fishing tackle. Sometimes he took his mother to concerts, lectures, and the theater. He also dated girls, on different occasions squiring Edith, Louise, and Kitty to the Powers Block, Rochester's skyscraper, where a dime bought a ride on the steam-powered "vertical railway" to the rooftop, seven dizzying stories up.

In 1877, when he was 23, George planned a vacation in Santo Domingo. A friend suggested that he buy a camera to record the trip in pictures. It sounded like a good idea, so he invested $94.36 in a stereoscopic camera "about the size of a soap box," along with a "packhorse load" of necessary paraphernalia. He never got to Santo Domingo, but he found his destiny—to simplify picture taking and "make it foolproof for the lazy, casual millions." He was not drawn, he admitted later, by the adventure of research. He smelled a money-maker.

His social life was now forgotten as he labored late into the night, experimenting, cooking chemical emulsions on his mother's kitchen stove. By the turn of the century, the world was snapping pictures with nifty little Kodak box cameras, and George Eastman controlled the bellwether corporation of a vast new Rochester industry. He established the Eastman Kodak Company on four basic principles that, if they weren't just a little ahead of their time, were ahead of his competitors: mass production, low product prices, foreign as well as domestic distribution, and extensive advertising. Just as important was his understanding of technology; he assumed that improvements would be coming rapidly, and bought up every photography patent that appeared to be necessary to the development of the business.

He realized, however, that he could not reach the top all by himself. At the office he was an exacting, impartial executive who ran the tautest of ships. Yet, in an age of sweatshop exploitation, he cut his employees' workweek from 54 to 49½ hours and gave his workers lunchrooms, a medical department, night classes, and a savings system, not to mention a welfare fund, life insurance, and a profit-sharing program to "do something for men who have grown old in our service."

He was generous and considerate in his private life, too. But his drive to the top hadn't left much time for developing social graces. He did all the proper things, but a man cursed with an undemonstrative nature was bound to be tagged with a reputation for coldness—especially if he was incapable of engaging in small talk. He did have close friendships with several women, almost always younger and safely married, but his mother was really the center of his emotional life. He went to her bedroom to kiss her good-bye every morning, and she sent him his lunch every day. Yet when he announced triumphantly that he had made his first million, she merely said, "That's nice, George," and dropped nary a stitch in

Kodak No. 1, patented by Eastman in 1888, shows its roll-film heart and circular images.

Eastman (with camera) visits a familiar-sounding Sudanese village in 1928.

The 1900 Brownie was for children.

charities, medical clinics, and the city of Rochester. He funded programs to remove the tonsils and fix the crooked teeth of Rochester's children.

He also hosted Sunday evening dinners and musicales, where two organs and the Kilbourn String Quartet entertained. Hardly intimate—a hundred guests was the norm—the parties were usually stiff and formal. Better, both for him and his friends, were the camping trips and trophy-hunting safaris, during which he proved himself a crack shot and a good cook.

He liked to eat breakfast with his personal organist, Harold Gleason, playing in the background. He was much taken by the "Marche Romaine," a lusty piece from a Gounod opera, calling it his "funeral march." While the last thunderous chord still echoed one morning, he shouted, "We'll give 'em hell when they carry me out the front door!"

That was in 1930, when he began to suffer from a progressive spinal disease. On March 14, 1932, he signed his will in his bedroom. He smoked a cigarette—Lucky Strike was his brand—in his black-and-gold holder and then penned a note on a sheet of lined yellow paper. It was as if he had rehearsed the scene in his mind many times. He stubbed the cigarette, capped his pen, removed his glasses and the skull cap he wore at home, lay down on the bed, folded a wet towel over his chest to prevent powder burns, and shot himself in the heart with a Luger automatic. The note, placed on the night table next to his mother's sewing basket, said: "To my friends My work is done why wait? GE."

A sad way to go, perhaps, but Harold Gleason provided a heroic note. At the end of the funeral service, as the casket was carried from the church, he launched into the triumphal "Marche Romaine" with full organ. "The effect," said Gleason, "was thrilling."

James A. Cox

the socks she was knitting for him.

"It came too late for me to enjoy," Maria confided to her nurse, and felt the same about the 37-room mansion George built for her. She died in 1907, and he spent the last 25 years of his life alone in the big house, surrounded by works of art, flower gardens, and unoccupied bedrooms labeled A, B, C, D, and so on. But he didn't lack things to do. "If a man has wealth," he said, "he has to make a choice, because there is the money heaping up. . . . It is more fun to give money than to will it."

By that definition, he had a ball. He gave an unprecedented $100,000,000 to universities, the performing arts,

Employees in Eastman's factory print by sunlight; holders contain developed negatives and photographic paper.

Camera and palette simultaneously capture a dancer's grace in a 1914 autochrome, the first practical color process. Introduced by the Lumière brothers in 1907, it added color to photos through a mosaic of potato-starch grains (left). Dyed red-orange, green, and violet, the grains were packed four million to the square inch. Finely powdered charcoal filled in between. Light-sensitive chemicals coated the grains, which filtered colors as light passed through. The developed plate produced a negative; exposed to light and redeveloped, it became a transparency dotted with specks of color far denser than those of a modern TV image. Although autochrome demanded long exposures and created dark plates, it remained popular until the 1930s. By then, another process—Kodachrome—was on the horizon.

had to be developed, fixed, and washed before the collodion dried.

The result of this wet-plate process was a glass negative. By harking back to Talbot's negative-positive process, Archer used the glass to print copies of the image on paper. He and other early users placed the glass against a dark background, essentially making his new wet-plate image into something that was hardly more than a glass daguerreotype.

One version of the wet plate became known as the tintype. The plate was not tin but a thin sheet of iron coated with black varnish. Using a version of the collodion process that Archer had pioneered on glass, tintypers produced within ten minutes portraits that were cheaper and more durable than daguerreotypes.

For half a century, tintypes would enshrine the faces of America: the wide-eyed young man off to the Civil War or the goldfields; the somber sweethearts left behind; the minstrels and preachers, the unsmiling children, a nimbus of movement blurring their heads; and the mothers, clutching their babies, stiffly portraying the madonnas of a new age.

The American land also became the subject of photographers romanticized as Pilgrims of the Sun, pioneers who headed West not to homestead but to record a pristine landscape. In wagons and on muleback they carried the usual provisions, along with the bulky apparatus of their art—huge cameras, bottles of chemicals, stacks of glass plates, lenses, a darkroom tent.

Unlike the tintypers, the pilgrims used the developed wet plates as negatives, financing their treks by selling prints of the wonders of the West. Photographic paper was typically coated with a mixture of development chemicals and egg albumen and then toned with gold chloride to rid images of what was called a cheesy color. Increasing demand for printing paper gave chickens an unexpected role in photographic history. One German company alone was cracking 60,000 eggs a day.

The pilgrims' images also found their way onto stereoscopic cards on which were printed two small photographs taken from slightly different points. Stereo fans inserted a card into a hand-held stereoscope and saw a three-dimensional view of Yosemite, Paris, or, sometimes, portraits of the family taken by local photographers.

The popularity of parlor stereoscopes showed a desire for images that reflected the reality of nature, including color. Touch-up artists supplied the auburn hair and the rosy cheeks until that day, promised by theorists, when color would be an integral part of the photograph.

The first experimental demonstration of color came in 1861, when

Austrian dancers in a 1938 NATIONAL GEOGRAPHIC magazine whirl to colors tuned by violinist Leopold Godowsky and pianist Leopold Mannes (upper). Blending talents in chemistry and music, they perfected the layered-emulsion process that led to Kodachrome film in 1935. In Kodak labs, legend had them whistling classics to time their experiments. Kodachrome's colors (left) are dyes—cyan (blue + green), magenta (blue + red), yellow (red + green)—introduced, layer by layer, during processing. Each of the three emulsions is sensitive to one of light's basic components. The layer closest to the film base responds to red; next comes green, then blue-violet. The separate images are developed and made into dye-formed positives. When superimposed over light, they join to create a full-color image.

A kicker's toe sinks into a football in a historic photograph that splits one of the seconds of 1938. High-speed light freezes the motion. Harold E. Edgerton, inventor of the electronic flash, or strobe, crouches (above) at a camera admitting the pulse of light during the shutter's relatively sluggish opening. The kick, one of the first strobe color photographs, added another moment to Edgerton's time-stopping gallery: a speeding bullet, splashing milk, a drop of water. Strobes—from the Greek for "whirling"—expose on film instants of motion too swift for the eye. A wink at a 40th of a second is glacial compared to the electronic flash's millionth.

James Clerk Maxwell, a young Scots physicist, produced separate red, green, and blue negatives of a tartan ribbon. He then made positive transparencies and projected them onto a screen using three lanterns that stood behind vessels filled with colored liquid. He superimposed the projected red, green, and blue images—and there on the wall was the tartan ribbon, the first, though fleeting, color photograph.

The experiment ushered in the theory of additive color photography, in which red, green, and blue images are added together to produce a realistically colored photograph. Inventors produced ingenious cameras and projectors to split and reassemble colors. But the systems were unwieldy and did not win wide popularity until 1907, when Auguste and Louis Lumière of France put their autochrome plates on the market. The brothers' secret was a layer of starch grains dyed red-orange, green, and violet. The layer was varnished and then coated with a black-and-white emulsion. After exposure to light, the developed image was reversed to produce a positive. The result was a kind of pointillism, in which tiny dots of different colors merged to form the impression of a full-colored image.

While color photography began a slow progress toward sharp natural hues, black-and-white photography got faster and easier in the decades around the turn of the century. Wet plates gave way to dry plates that did not need bulky equipment for on-the-scene developing. Cameras, though still shackled to unwieldy glass plates, got shutters and better lenses. The ranks of amateur photographers continued to grow.

One of the amateurs was George Eastman, a disgruntled bank employee in Rochester, New York. He knew there would be even more amateur photographers if they did not have to contend with heavy, breakable glass plates. They needed flexible film.

In 1880 Eastman got a U.S. patent on a machine for coating the emulsion on glass plates. A year later Eastman quit his job. Soon running a prosperous photography business, he invested his capital and talent in plans to make the snapshot his worldwide commodity.

In 1888 he marketed a $25 camera known simply as the Kodak. It weighed a pound and a half and held a spool of paper-backed film long enough for taking a hundred pictures. The amateur snapped a picture, rolled to the next exposure, and pulled a string to wind the shutter mechanism and prepare the camera for the next shot. When the film was finished, the camera was mailed back to the factory. After developing and printing the film, the factory returned the camera and a set of contact prints and, for $10, loaded the camera with another roll of film.

The Kodak put photography in everybody's hands. As the instruction booklet said, "This is the essence of photography and the greatest improvement of them all; for where the practice of the art was formerly confined to those who could give it study and time and room, it is now feasible for *every body.*"

In 1889 Eastman put the emulsion on a recently invented substance called celluloid and soon afterward introduced a flexible roll of film that could be removed from the camera. No longer did the camera have to be returned to the factory; users could load cameras anywhere without

Edwin Land

E dwin Herbert Land is a creator of images. Glare-free, polarized images; instant photographic images; images of Edwin Herbert Land. For photographic images he will be most remembered; for the images of himself—scientist, philosopher, businessman, magician—perhaps most admired. But the polarized image is what started it all.

"If you can think of it, you can do it," Land once said. He thought of Polaroid, the plasticized sheet that was his first invention, as he strolled down Broadway one night in 1927, an 18-year-old freshman visiting from Harvard. His eyes darted from theater marquees to neon billboards to headlights; he was stunned by the glare. He knew that he could eliminate it if he could channel light waves, which normally vibrate in a multitude of planes, through a filter that would direct them all into one vibrating plane.

Light had been polarized before, through various crystals. But the idea of a sheet of polarizing material that might fit over a headlight or windshield was original. Land stayed in New York, found a physics laboratory at Columbia he could climb into at night through an unlocked window, and by 1929 had developed a plastic in which microscopic polarizing crystals were aligned in parallel rows. He called it Polaroid.

Land never graduated from Harvard, or from any other university, although he has received honorary degrees from 15. In 1937 he founded the Polaroid Corporation in Cambridge, Massachusetts, and his plastic sheets became Polaroid Day Glasses.

With the birth of the corporation, Land—its chairman, president, and research director—began to create a public image for himself. He was a most quotable man. The progressive

Land unveils an instant image in a 1947 demonstration (upper). Crossed Polaroid filters trap light's glare (lower).

and benevolent patriarch of Polaroid: "The task of the research bureaucracy is to lend a sensitive ear to the whispers of shrewd intuition." The brilliant, driven scientist-inventor: "I'm an addict—addicted to at least one good experiment a day." The philosopher: "Every good picture we take . . . should make our lives that much bigger. Photography is an illustration of the use of technology not to estrange, but to reveal and unite people." The magician: "A good scientist has a charlatan's drive. His urge is to make magic."

That was the role Land seemed to like best. "A Genius and His Magic Camera," announced a cover of *Life*

magazine in 1972. There was the face of the man who claimed to hate personal publicity, smiling behind the lens of his latest invention, the SX-70.

Land's first instant camera had debuted on February 21, 1947. During the winter meeting of the Optical Society of America, Land took two instant pictures of himself. The result of the first 60-second developing and printing process took the audience by surprise; the second picture was recorded by every photographer there and published in newspapers worldwide.

Land actually invented instant photography in 1943, when he, his wife, Helen, and their three-year-old daughter, Jennifer, were vacationing in Santa Fe. During a day of sight-seeing Jennifer asked why she couldn't see their snapshots right away. Land, startled, wondered the same thing. Within an hour, he had envisioned a camera containing two rolls of photographic paper and an individual pod of developing chemicals for each picture. The photographer would expose the image on negative paper, pull it down to meet a roll of positive paper, then pull both through a pair of rollers that would burst one pod, spreading the chemicals evenly throughout the negative-positive sandwich. Three years later he peeled apart that sensational sandwich, revealing to the dazzled Optical Society the now famous sepia image of his own darkly handsome face.

When Land retired from Polaroid in 1982, he held 533 U.S. patents in photography and related fields, more patents than any inventor besides Thomas Edison. Incredibly, many of Land's inventions took shape in a matter of hours. He views his work as eminently important, the instant camera as "a magic device . . . an invaluable instrument for discernment of [our] prehistoric bonds to each other."

For all that, Edwin Land claims he might never have invented that camera if it hadn't been for Jennifer's simple question.

Lynn Addison Yorke

A 1976 still life on Polacolor film by Marie Cosindas, who called the instant process "the darkroom in my hand."

Galloping into photographic history, a freeze-framed horse answers a question no eye could answer: Does a running horse ever have all four feet off the ground? The photos came from Eadweard Muybridge (opposite), who switched from photographing landscapes to freezing time. In the 1870s, with Leland Stanford, former governor of California, as his patron, Muybridge built equipment to answer the four-fleet-feet question—and, by legend, settle a $25,000 bet. At a special running track in Palo Alto (upper), he set up 24 cameras whose electromagnetic shutters were tripped when a horse broke threads strung across the track. Lines on a fence provided a time-motion gauge. His zoopraxiscope (right) projected glass-plate copies of sequence photos, giving an illusion of motion and foreshadowing the system that would put movies on a screen.

a darkroom. In 1900 Eastman brought out the first Brownie, designed for children. Photography became even more simplified for the amateur and more profitable for Eastman.

The camera, with faster lenses and films, ventured indoors, its light generated by exploding magnesium powder. Soon, electrical inventions contributed to photographic lighting, beginning with the safe, silent flashbulb and continuing through electronic flash equipment.

As photography entered its second century in the 1940s, there seemed little left to invent. But Edwin Land, an American scientist who had already tamed light with his invention of a glare-reducing material called Polaroid, decided that people should not have to wait to get their photographs developed. After spending three years on the problem, in February 1947 he unveiled his one-step dry process for producing a photograph one minute after the photographer clicked the shutter. The following year, when the Polaroid Land camera went on sale, he began

research on instant color film, which would enter the market in 1963.

Land, second only to Thomas Alva Edison as a U. S. patent holder (the score: 1,093 to 535), shared another accomplishment with Edison. Their inventions gave new dimensions to motion pictures. During the brief 3-D movie rage of the 1950s, moviegoers saw the action through disposable, Land-produced Polaroid glasses. Edison's involvement with the movies was more lasting. He was one of many inventors who transformed photographic imagery from still life to a moving, talking spectacle. Some of the inventions even preceded photography.

The ancestors of the movie projector were magic lanterns that used lenses and sources of light to project images on the wall. Not until the arrival of photography, however, did inventors get the raw materials for the creation of a powerful new medium that would not only capture reality but also distort it. Illusion is the essence of motion pictures.

Motion picture inventions stem from that reality of illusion set forth in 1824 by Peter Mark Roget (better known for his thesaurus of the English language). In his paper "The Persistence of Vision with Regard to Moving Objects," he observed that the human eye and mind retain an apparent image for a moment after the image has vanished. This is why

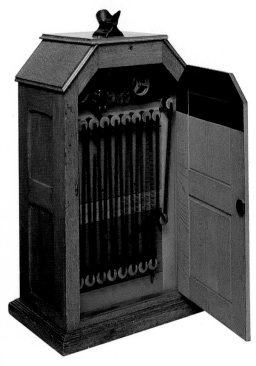

we can simultaneously see both the head and tail of a spinning coin.

Roget's concept fascinated tinkerers. One of the first gadgets inspired by his discovery was the zoetrope. This novelty bestowed the illusion of movement on a sequence of revolving sketches viewed through slots in the side of a spinning drum. Later devices portrayed motion through photographs. But only a sequence of high-speed photographs could produce a believable illusion of animation.

Those photographs would come first from British-born Eadweard Muybridge, a Pilgrim of the Sun who, around 1878, began building an apparatus for photographing moving objects. Muybridge, using electrically triggered camera shutters, succeeded in stopping, in 1,000th-of-a-second exposures, the legs of galloping horses and human runners.

Muybridge packed sequential photographs into his zoopraxiscope, a projector that displayed his repertory of motion on a screen, anticipating the system that would be used for showing movies. On a lecture tour of Europe in 1881, he demonstrated the projector to a French scientist, Étienne-Jules Marey, who was entranced by the idea of freezing motion. Six months after talking to Muybridge, Marey built his photographic rifle—a camera lens mounted in the barrel of a gun and operated by the trigger. By exposing a dozen images a second on a revolving plate, the rifle could stop the wings of a bird in flight. Muybridge and Marey, each on his own path, were near the threshold between stop-action images and motion pictures.

Now photographic history itself suddenly begins to look like a series of rapidly moving scenes. *February 1888:* Muybridge and Edison meet and talk inconclusively about linking the zoopraxiscope to the phonograph, which Edison had invented in 1877.

October 1888: Marey demonstrates a camera capable of taking up to 60 photographs a second. That same month Edison takes steps to patent a projector he calls the Kinetoscope.

August 1889: George Eastman puts celluloid roll film on the market. Marey buys some, makes a series of photos of a running horse, and tries unsuccessfully to project a moving strip of the images onto a screen. Edison meets Marey in Paris, sees his roll-film camera.

September 1889: William Dickson, an Edison employee, writes to Eastman and puts in a special order for roll film that is extremely narrow and 54 feet long. Dickson begins working on a profoundly simple idea that would have solved the problem of Marey's camera. To move roll film rapidly through a camera and later through a projector,

Thomas Edison, seeing movies as a novelty to be sneezed at, in the 1890s launched a peep-show industry that recorded such dramas as a very long a-a-c-h-h-o-o-o! (opposite) and The Execution of Mary Queen of Scots, *in which the royal head rolls. Edison's neophyte moviemakers packed their art into the Kinetoscope (left). In peep parlors (upper), when patrons dropped a nickel into a slot, a* battery-operated motor moved a strip *of pictures on transparent flexible film—itself a new invention. As the celluloid film ran past a magnifying lens and an electric lamp at the rate of about 40 frames a second, the viewer got the appearance of motion from the flickering images. Peep shows, sometimes as short as 20 seconds, faded out when movie fans began craning their necks in a new way: toward a screen.*

174

Stars of cinema invention, Louis Lumière (peering at an image) and his brother, Auguste, work in their laboratory in Lyon, France. The Lumières lived up to their name—"light" in French—by better illuminating movie film so that it could be shown on a screen. After seeing Edison's Kinetoscope, they envisioned a machine, the cinématographe *(lower), that produced enough light for projection.*

The cinématographe *solved the problem of motion picture continuity by being both camera and projector. The brothers opened a movie theater in a Paris café basement in 1895, and audiences were soon applauding such short films as* The Sprinkler Sprinkled *(opposite) and* Baby's Breakfast.

Dickson perforates the sides of the film and makes a sprocket system—the enduring mechanism of cinema photography.

On May 20, 1891, members of the National Federation of Women's Clubs are touring Edison's laboratory in West Orange, New Jersey. They are invited to peek through a hole on the top of a small wood box. "As they looked through this hole," says an account of the visit, "they saw the picture of a man. It bowed and smiled and waved its hands and took off its hat with the most perfect naturalness and grace." The women have just seen a movie.

In the spring of 1894, Edison's Kinetoscope Company began shipping the wood boxes to peep-show parlors throughout the country. For a nickel, a patron of the new art form put an eye to the flickering movements of a Spanish dancer, a wrestling match, a comic pantomime, or whatever else Edison's neophyte moviemakers could dream up.

What unpredictably became the movie business began moving at a fast rate. The peep shows were appetizers for a public that had suddenly discovered a hunger for celluloid entertainment. Inventors in the United States, France, England, and Germany frantically worked on ways to move the movie from the peep show to the screen.

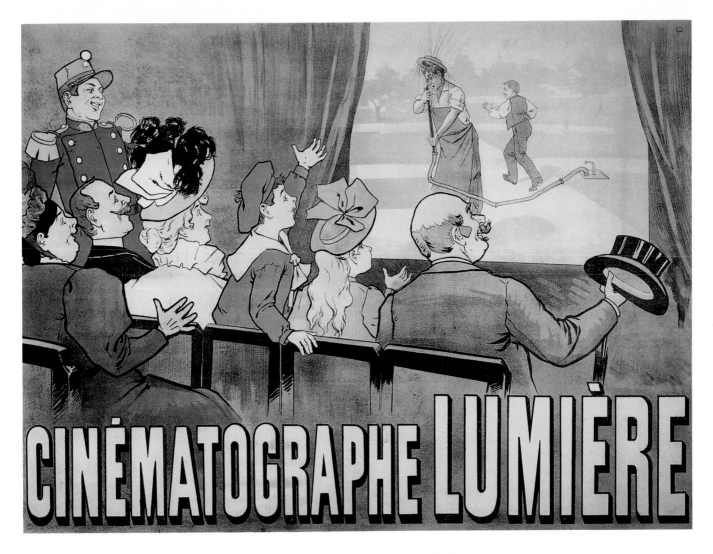

CINÉMATOGRAPHE LUMIÉRE

Edison backed another inventor's projector, the vitascope, which premiered in a New York music hall in April 1896. A spectator enthused about the dancers—"as if the girls were there on the stage"—and raved about the breaking ocean waves: "Some of the people in the front rows seemed to be afraid they were going to get wet, and looked about to see where they could run to, in case the waves came too close."

Within a year the vitascope had been eclipsed in New York by a Parisian import, the *cinématographe,* a combination movie camera and projector devised by the Lumière brothers, who insisted that their work on color film was far more important than this novelty. The fare included such epics as *The Family Tea Table* and *Baby and the Fish Bowl.* But audiences cheered the "living picture" that appeared on the screen, bigger than life and somehow more exciting than actors and actresses strutting on a mere stage.

Soon the peep show was supplanted by what Americans called their new five-cent theater—the nickelodeon. Entrepreneurs rented vacant stores, brought in some seats and a piano, hung up a canvas screen, and collected a niagara of nickels. By 1910, in the United States alone there were about 10,000 nickelodeons, and enough 10- to 15-minute

movies were being ground out for a new bill daily.

And each day there seemed to be a new, often dubious invention. Some inventors shuttled between their workshops and courtrooms, where dozens of patent-infringement battles droned on. Movies were so bad that vaudeville house managers used them as "chasers" to send the audience scurrying out of the theater to make room for the next show. By the 1920s, moviemakers were desperately seeking a way to hold their jaded, disenchanted fans. The call went out for inventions that would add more realism to the illusions on the screen.

Real life had color and voice. To add those realities to the movies, inventors in both the United States and Europe came forth with ingenious devices that had their moments of success and then vanished. Complex color and sound systems overwhelmed the typical movie house, whose technological resources amounted to hardly more than a balky projector and a piano.

If movies did not talk, money did. In 1927 Warner Brothers, a small, financially shaky movie company, risked bankruptcy to produce *The Jazz Singer,* a sound film that originally was to have had only music and song. But the star, Al Jolson, after finishing a song, blurted out a line of dialogue: "Wait a minute! Wait a minute! You ain't heard nothin' yet!" Jolson's prophetic words launched the talkies. Within the next three years Warner Brothers bought 250 theaters, and company assets increased from $16,000,000 to $230,000,000.

The first feature-length complete talkie, *Lights of New York,* came out in 1928. It cost Warner Brothers $23,000 to make and earned $1,252,000. No one could now doubt the draw of talkies. Undaunted by the Great Depression, movie theaters began borrowing money to buy sound equipment. By 1930 nearly half the movie houses in America—and one out of every three theaters in the world—could show talkies.

From the beginning, the making of movies depended upon teams of people, arrays of machines. One word summed it up: studio. The Black Maria (opposite), the world's first motion picture studio, opened in 1893 at Edison's laboratory in West Orange, New Jersey. Built of tar-paper-covered pine to ease weight, it could be turned to follow the sun. The Black Maria was hardly more than a stage, with a table-mounted camera that moved forward for close-ups. Within four decades, movies went from silence to a roar: An intrepid cameraman and sound man record the Metro-Goldwyn-Mayer lion. MGM, Home of the Stars, headed for the pinnacle of a Hollywood that had discovered sound. During the 1930s and '40s, when movie houses sold 80 million tickets a week, MGM made more films than any other studio.

In epics like 1916's Intolerance (left), creative photography came from better lenses, dependable film, artificial light, and such simple inventions as tracks to move cameras. The new tools made magic in the hands of D. W. Griffith, director of Intolerance and forerunner of movie artists who would make film live. The camera, once only an observer of events, became part of the action, roving for panoramic shots, lunging in for close-ups. Griffith learned his art first as an actor. In Rescued from an Eagle's Nest (below), a 1907 film, he battled an early special effect—a stuffed bird playing an eagle that carried off a baby.

Sound and color burst onto the screen because, behind it, moviemakers demanded new inventions to sell more tickets. When the pioneering talkie The Jazz Singer *(opposite) spoke in 1927, only about a hundred movie houses were wired for sound. But sound movies quickly flourished, and the film's near-bankrupt studio, Warner Brothers, won its gamble. Talkies inspired scores of inventions: sound-on-film, electrically driven cameras that synchronized sight and sound, improved projectors. After years of flirting with color, moviemakers backed Technicolor, whose two-color process became three-color "brighter than life" in the 1930s. The new process used three separate films in a beam-splitting camera. The 1939 classic* The Wizard of Oz *(above) was one of many films that dramatized Hollywood's switch to color by startling audiences with a sudden shift from black-and-white to Technicolor.*

The sound of *The Jazz Singer* came from a record synchronized with the film. The Western Electric system that Warner Brothers used soon gave way to one that put sound directly on film. Using a principle developed by inventor Lee De Forest, a microphone converted sound to electricity. Variations in the current were indicated by the changing intensity of a lamp. The camera filming the scene had a two-track film, one for sight and the other for sound. The lamp's pattern of light was picked up by the sound track. When the movie was projected, the squiggling line of the sound track was electronically converted back to sound and amplified through a loudspeaker array in the theater.

The inventions associated with sound had an unexpected effect on the artistry of movies. Noisy cameras had to be encased in soundproof housings, limiting mobility. A new technician, the sound engineer, challenged the tyranny of temperamental directors who were used to yelling at the cast during filming. While directors silently fumed on the now sound-conscious set, the actors spoke their lines in ways that the director might not want. For it was the engineer who decided where to hide cumbersome microphones and their serpentine cables.

So it would be this newcomer, not the director, who would decide where an actor turned (toward the potted plant or the star's bosom, for instance) and when he spoke—or perhaps shouted—his lines. Soon, though, shouting directors learned to live with the new techniques. They photographed scenes without sound and, thanks to other inventions, dubbed the sound in later.

Color arrived erratically, first as hand-tinted frames, then as variations on the still camera's early three-filter process. Kodacolor film, introduced by the Eastman Kodak Company in 1928, was intended for amateurs, but moviemakers adapted it with moderate success. Finally came Technicolor, which began its career in 1917 in a railway car that rolled from Boston, where the bulky color-filming equipment was loaded aboard, to Jacksonville, Florida, where a pioneer color movie, *The Gulf Between,* was being made.

The system was so complicated that its originator and promoter, Herbert Kalmus, said it needed a projector operator "who was a cross between a college professor and an acrobat." Technicolor evolved from that primitive two-color technology to a full-spectrum system that used an unwieldy three-film camera, and finally to a simplified process that used a variation of Kodachrome film.

For the advent of color there was no epochal film like *The Jazz Singer.* Color flitted in and out of movies, teasing the moviegoer with hints of tints to come. We sat there in the black-and-white darkness, awaiting the rainbow. Then one day it came, as suddenly and unexpectedly as so many things come to the spectators of technology.

Like all images that really matter, this one would stay always in the newsreels of our minds: There, one moment, was Dorothy in a black-and-white Kansas tornado. A gray door opened—and the next moment she was somewhere over the rainbow in the brilliant, sparkling colors of Oz. In that dazzling darkness we knew that we all had been taken to a new place and, unlike Dorothy, we would never go back.

NOW PLAYING
DIRECTION WARNER BROS.
WARNER
Theatre
WARNER BROS.
AL
IN
WITH
TWICE DAILY
Sunday Matinees
WARNERS
AL JOLSON in
"THE JAZZ SINGER"
WARNERS'
NOW PLAYING

Messages by Wireless

By Fred Strebeigh

I WILL NOT TALK RADIO, read the sign on the door of the laboratory of the old inventor, Thomas Edison, TO ANYONE. The year was 1921. And everyone, all America, even the young man knocking at Mr. Edison's door, seemed to be talking radios. And building radios. And buying radios, if they could find them.

Less than a year before, the United States had no such thing as a licensed commercial broadcasting station. Within a year and a half, it would have more than 500. Everyone felt the thrill of voice and song emerging out of thin air. But Thomas Edison remained unthrilled. Seventy-four years old, crusty and persnickety, the holder of patents for the light bulb and the phonograph and more other devices than anyone else in history—Mr. Edison wished radio would leave him alone.

And now there came knocking at the door this young man, Tommy Cowan, who had gotten his first job with Mr. Edison but later switched over to Westinghouse, Edison's biggest rival. Today, young Tommy wanted to borrow one of Mr. Edison's phonographs. And he wanted to hoist it up to a rooftop not far from the Edison lab in West Orange, New Jersey, because tomorrow, up there, Westinghouse was starting another accursed radio station.

Mr. Edison lent the phonograph and a few records as well. But he was none too happy, Tommy gathered—perhaps because radio, the force that began as a little spark but now was sending waves around the world, had almost been discovered by Mr. Edison himself.

All that was years ago. In 1875 Thomas Edison had been experimenting with electric telegraph equipment when he noticed sparks leaping from iron placed near a magnet. He found he could make the sparks leap from metal pipes across the room. Energy seemed to be transmitted through space, without wires. In his laboratory notes, Edison predicted that he had found a *"true unknown force."* He called it "etheric force." Not for years would it be known as radio waves or, more precisely, electromagnetic radiation. But though it had no name, it had a history, recorded partly in odd events and partly in theory.

One such event occurred in 1842, when the American physicist Joseph Henry magnetized ordinary needles by placing them within a couple of hundred feet of electrically charged wires. A still odder event occurred in the 1860s, when an American dentist, Mahlon Loomis, sent signals from one mountaintop to another. At the bottom of a long wire hanging from a kite, Loomis produced a series of electrical impulses. His associates, flying a similar kite 18 miles away, apparently detected impulses that coincided with those produced by Loomis.

Such events posed a theoretical problem. A basic physical law stated that an electrical charge could be neither created nor destroyed. Yet, in

Henry's experiment, an electrical charge seemed to have "leaked" from the wires; at the same time a magnetic charge seemed to have been created 220 feet away. Similarly, in Loomis's experiment, an electrical charge appeared inexplicably in the kite wire flown by his associates.

In 1864 a Scottish physicist, James Clerk Maxwell, offered a theory to solve the problem. He offered his theory in the form of mathematical equations, but unfortunately he presented no physical model to help others understand those equations. Worse, his math suggested the unimaginable: Electricity and magnetism flow as one force, not just through conductors like wires but through open space. This flow takes the form of waves in what Maxwell called an "electromagnetic field." Maxwell's math implied that these electromagnetic waves travel at the velocity of light waves and, furthermore, that light constitutes merely one type of electromagnetic radiation—the best known type, since one of its best detectors is the human eye. Maxwell was correct. But since he offered all math and no physical model, his theory proved incomprehensible and major scientists rejected it.

Then, in 1875, young Thomas Edison sat watching as his etheric force generated electrical sparks in far corners of his lab. Although he was working to develop a wireless telegraph, he could not imagine how his new force could carry information to the far corners of the nation. For once in Edison's life, he had the spark but not the inspiration. The secret of radio escaped him.

Finally, in 1888, a young German physicist, Heinrich Hertz, announced a series of brilliant experiments. In a large lecture hall, Hertz had set up a simple transmitter and receiver. When the transmitter sparked, the receiver responded with a smaller spark, showing that some electrical energy had radiated across the room. Using techniques similar to those already devised for studying light, Hertz demonstrated that his transmitter created waves that moved through space at the speed of light. Later, Hertz went on to demonstrate that these waves could be manipulated as light could—bent by prismatic objects, bounced back by metal plates, or focused by parabolic reflectors.

The simplicity of Hertz's experiments led excited scientists around the world to replicate them. Journalists and lecturers spread the word. An article in London's influential *Fortnightly Review* proposed that Hertz's waves be combined with Morse code to produce "telegraphy without wires." To such practical uses, however, Hertz seemed oblivious. The stage was set, spectacularly, for the right entrepreneur.

Listening for radio emissions from black holes, exploding supernovae, and other such phenomena, New Mexico's Very Large Array—27 dish antennas like this one—reaches farther into space than any other radio telescope. Beginning as experiments with electric sparks in 1888, wireless inventions now provide instant global communication and even, as here, receive messages from the stars.

Marconi to father Neptune —
I can beat you out any time
if you will only give me a few more of these life boats —

Grant Wright N.Y.

Wireless telegraph operator David Sarnoff, later a radio and television tycoon, relays the story of the Titanic *disaster to a horrified public from a Marconi station in New York.*

Distress calls from the Titanic's *wireless operator around midnight on April 14, 1912, had revealed that the liner, gashed by an iceberg, was sinking fast. The* Carpathia *responded, racing from 58 miles away to rescue 705 people from lifeboats. The cartoon honors Marconi's invention of the radio link that saved these lives. For lack of more boats, 1,522 people froze or drowned in the icy waters. Many of them might have survived if the wireless operator on the* Californian, *only 20 miles distant, had heard the calls, but he had gone to bed. Congress hastily passed laws requiring adequate lifeboat space and a continuous radio watch on large passenger ships.*

That entrepreneur arrived in Britain from Italy early in 1896 carrying a black box. To British customs officials, the box looked like a terrorist device. They took it apart and broke its mechanism.

Undaunted, the young entrepreneur, 22-year-old Guglielmo Marconi, rebuilt his shattered kit—a simple wireless communication system comprising a transmitter, a receiver, and a copper-strip antenna. For the future of wireless, the man mattered more than the box he carried. The Marconi system offered only a few technical improvements on the work of others, such as Oliver Lodge in Britain, Aleksandr Popov in Russia, and Karl Ferdinand Braun in Germany. But Marconi the man offered something new. Unlike the other experimenters, he had a single goal—to devise a practical system for sending messages with Hertz's radiating waves or, as they would soon be known, radio waves.

By 1897 Marconi received the first patent ever granted for radio and helped organize a company to market his system. By 1901 he announced receipt of signals that spanned the Atlantic. By 1907 all the great transatlantic ocean liners carried Marconi radio equipment and telegraph operators, tapping out news to Marconi land stations as their ships sailed and, at times—as from aboard the *Titanic*—calling out in desperation for help as their ships went down.

Even as Marconi's dots and dashes multiplied, however, their decline was in the air. On Christmas Eve, 1906, at an experimental transmitting station on a Massachusetts coastal farm north of Plymouth harbor, Reginald Aubrey Fessenden prepared a gift for ships at sea. Fessenden believed that radio waves, which everyone from Hertz to Marconi had generated with sparks, could be created by whirling generators that produced high-frequency alternating current. When the sputter of sparks gave way to a high-speed whir, Fessenden believed, continuous waves could carry something new: the human voice.

Guglielmo Marconi

The family legend is probably too good to be true. On the day Guglielmo Marconi was born in Bologna, so the story goes, the servants flocked into the bedroom to admire the new baby. An old gardener cried, *"Che orrechi grandi ha!"*—"What big ears he has!"

"With these ears," his mother replied hotly, mixing Scotch-Irish fire and poetry, "he'll be able to hear the still, small voices of the air."

So he did, in a manner of speaking. And from then on, with one memorable exception many years later, the voices that filled the air were still and small no more.

From his birth in 1874 to early manhood, Guglielmo was molded by the temperamental differences of his parents. His father, Giuseppe, landowner and widower, was 35 when he fell in love with 18-year-old Annie Jameson, a Scotch-Irish songbird visiting Bologna to study *bel canto*. Giuseppe proposed. The whisky-distilling Jamesons, having sent the lass abroad to keep her from performing at London's Covent Garden (which nice girls didn't do), thundered their displeasure and forbade the marriage. Respectfully rebellious, the lovers waited until she was 21 and then eloped. A year later, Alfonso was born. Nine more years and Guglielmo arrived. Annie almost died in birthing him, and that trauma produced a strong mother-child relationship. Still young, she shielded her frail, blond-haired little boy from the harsh discipline of Giuseppe, now a middle-aged, lira-pinching, domestic *tiranno*.

She taught him English and the piano, and arranged for a tutor. Shy and withdrawn, Guglielmo often played hooky from his studies, drawn to the fascinating world of nature, to the solitary world of fishing, and to the mind-opening world he found in his father's

Guglielmo Marconi with his mother, Annie.

library—Greek mythology and history at first, then steam engines, chemistry, and Michael Faraday's classic lectures on electricity. When he did go to school, he fared poorly, mocked by his classmates and held up to ridicule by a teacher for his English-accented Italian. His shell of aloofness thickened.

Reading was never enough for him; he had to put what he read to the test. He constructed a working still. He dismantled a cousin's sewing machine to make a mechanical roasting spit, then put it back together when she wept. Inspired by Benjamin Franklin's *Life,* he shot high-voltage electricity into an elaborate contraption of string and dinner plates, with smashing results. Giuseppe, already irate over the way his son was wasting his life tinkering with foolish gadgets, was enraged.

He was just as mad when Guglielmo flunked out of secondary school, and then failed the entrance examinations to the Italian Naval Academy and the University of Bologna. A technical school in Leghorn was more to the boy's liking, and two things happened there that were to influence his whole life: A teacher introduced him to electrophysics, and a blind telegrapher

taught him how to send Morse code.

In 1894 Guglielmo Marconi's life changed forever; what had been an adolescent fascination now became, at age 20, a passion. On holiday in the Italian Alps, he read about Heinrich Hertz's experiments with the transmission of electromagnetic waves. Into his mind leaped the possibility of adapting the Hertzian waves to a system of wireless telegraphy. The idea, he said later, "was so simple in logic, that it seemed difficult for me to believe that no one else had thought of putting it into practice."

Simple in logic, but not easily proved. Earlier, indulgent mother had prevailed upon cantankerous father to allow eccentric son to set up a "laboratory" among the abandoned silkworm trays in their attic. Here, Guglielmo feverishly buried himself in his project, fearing only that someone else would beat him to the prize, in his mind a puerile porridge of fame, glory, financial gain, and adulation by pretty girls.

But there was nothing childish about his tenacity. For months he labored single-mindedly, rarely seen by his family, missing meals until the worried Annie took to leaving trays of food outside his door. But at last he shared with her a great moment. Late one summer night he roused Annie and led her up to the attic. As she watched, he bent over a telegraph key and tapped it. From the far end of the attic a bell tinkled. Between key and bell, no wire—only air. Wireless telegraphy, ancestor of crystal radio sets, Atwater Kent radios, television, radar, and maybe even Star Wars, had been born, sired by a 20-year-old loner who wanted to justify his mother's love and prove to his father that he was worth something after all.

That was only the first tottering step, of course. From the attic he moved to the garden, to open fields, to hillsides, to the seacoast, refining his equipment, studying, stretching by feet and then by miles the distance he could throw the signal, and ignoring the experts who dogmatically declared that it

A museum in Rome displays a transmitter circuit in the reconstructed laboratory from Marconi's yacht *Elettra*.

190

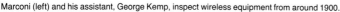
Marconi (left) and his assistant, George Kemp, inspect wireless equipment from around 1900.

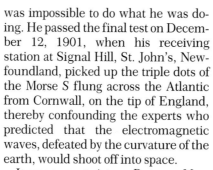

was impossible to do what he was doing. He passed the final test on December 12, 1901, when his receiving station at Signal Hill, St. John's, Newfoundland, picked up the triple dots of the Morse S flung across the Atlantic from Cornwall, on the tip of England, thereby confounding the experts who predicted that the electromagnetic waves, defeated by the curvature of the earth, would shoot off into space.

It was a great victory. But now Marconi the scientist had to share the stage with Marconi the entrepreneur. The latter, dapper, punctual, the antithesis of the public's conception of a serious man of science, formed companies and struck deals, while the former patented inventions that improved signal transmission and reception and, in 1909, accepted the Nobel Prize for physics. Marconi the man also showed up. He married twice, fathered several children, had several affairs, and bought a 220-foot yacht, the *Elettra,* that served as a floating laboratory for scientific experiments as well as for shipboard parties, to which his first wife, Beatrice, was not always invited.

Rich and famous, laden with international honors, Marconi still spoke of his "higher destiny" and experimented with shortwaves and microwaves. But his most productive days were behind him, even as he continued to bask in the role of the great inventor. Bad health and melancholia dogged him as he grew older, and on July 20, 1937, he died in Rome of a heart attack.

The next day, in tribute, wireless operators all over the world shut down their transmitters for two minutes—a silence broken only by the still, small voices of the air.

James A. Cox

The Signal Hill team at St. John's, Newfoundland, battles a gale to launch a kite-borne antenna for the transatlantic radio test on December 12, 1901.

Marconi (third from left) and staff at the Canadian wireless station opened at Glace Bay, Cape Breton Island, in 1902.

Dame Nellie Melba's clear soprano soars across European and Atlantic airwaves from the Marconi wireless factory in Chelmsford, England (opposite). On June 15, 1920, the Australian prima donna gave Britain's first advertised program of entertainment on the radio—until then used mostly as a news service. High quality broadcasting came from the invention of the triode vacuum tube (above), which amplified weak electronic signals. Patented by American Lee De Forest (top) in 1907, it revolutionized radio and telephone equipment and provided the key to the development of radar, television, and computers.

On that Christmas Eve, wireless operators on American Navy and merchant ships off the Atlantic shore sat listening as always at their headphones. They were used to hearing DASH-DOT-DASH, the flat blips of Morse code. Then, at Fessenden's station, an alternator whirred, a microphone hummed. Up and down the Atlantic, operators strained to hear. Headphones ceased their relentless blipping. A man spoke. A violin trilled. The headphones played "O Holy Night."

Fessenden's Christmas Eve performance brought no financial triumph. His wireless system went unsold. His company went bankrupt. His alternator-transmitter, which General Electric had built to his specifications and later improved, went on to make money for GE. But Fessenden had introduced to radio the sound of the human voice and, by most accounts, the concept of the broadcast—wireless transmission intended for a wide audience.

Both were premature. Voice broadcasting demanded more sophisticated transmitters and receivers. This equipment would come, eventually, from a single complex invention—the triode, which had its origin in another of Thomas Edison's creations, the light bulb. What the triode provided was amplification, the ability to turn small signals into large ones. But it would take another decade and a world war before triodes could emerge as the most numerous and significant of all the little glowing tubes in radio history.

In 1904 the near-inventor of the triode, John Ambrose Fleming of England, was trying to develop an improved wireless receiver. He hoped to refine the detector—the device that sensed the reception of radio waves. Fleming thought back two decades to his research on a puzzling electrical effect first discovered by Edison—a tendency for dark particles to smudge the inner surface of glass light bulbs as current flows through in a single direction. Fleming pulled from a cupboard an old bulb, specially fitted with two electrodes, which he had used to study the puzzling "Edison effect." Slightly modified and attached to a simple radio receiving system, the bulb proved capable of turning radio oscillations into a detectable flow of direct current. With further modification, this two-electrode bulb or tube became known as the diode. Though it would become integral to a radio's power supply, Fleming's diode unfortunately proved no better than existing detectors.

Within two years, an American, Lee De Forest, seeking not a detector but an amplifier, developed a diode virtually identical to Fleming's. He then added a third electrode, creating a triode. When attached to a radio receiver, the new tube boosted the strength of weak signals.

For years, however, sidetracked by business matters and radiotelephone experiments, De Forest did nothing to further develop the triode. His greatest triumphs came instead from broadcasts: phonograph music in 1908 from the Eiffel Tower in Paris, arias sung by Enrico Caruso from New York's Metropolitan Opera House in 1910. Only later would the full capabilities of his triode emerge in circuits that produced sensitive reception and powerfully amplified transmission.

Then came World War I. Desperate for clear radio communication on land and sea, the United States government invoked its war powers.

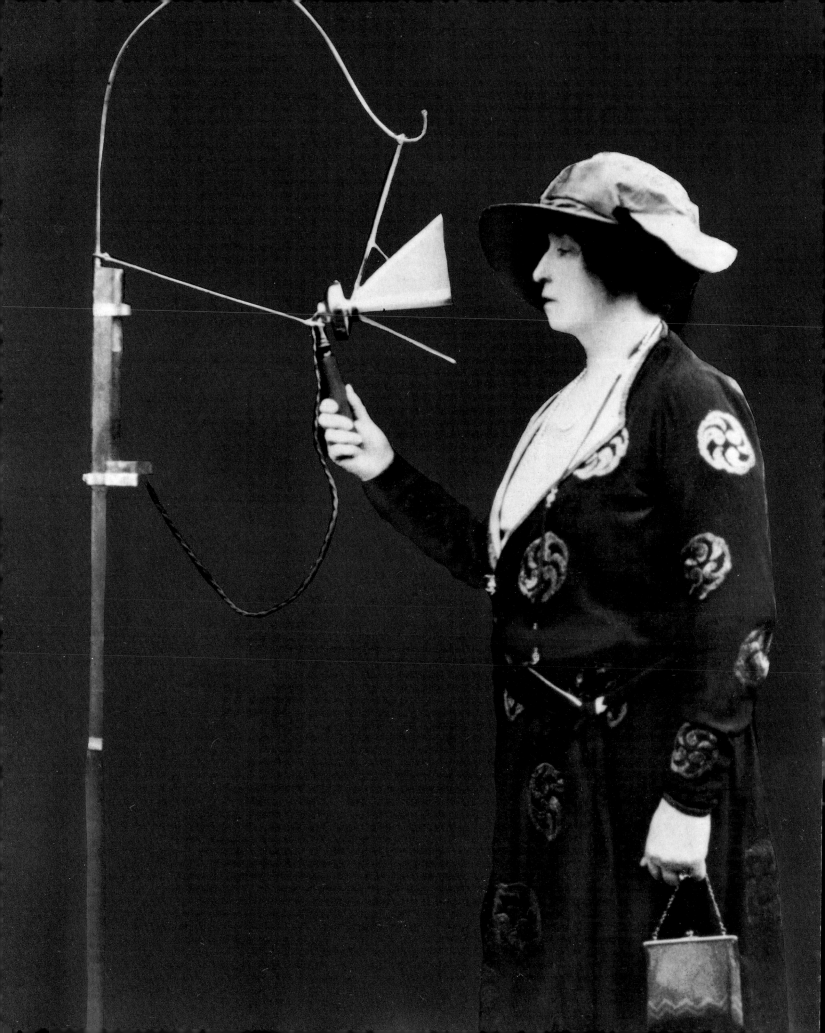

It banned civilian radio, suspended all exclusive manufacturing rights on patented material, and set American industry to building military radios at full speed. Factories that had made light bulbs now made triodes by the thousands. Navy wireless schools trained ten thousand men. Radio gear for communicating from ship to ship and regiment to regiment became lighter, clearer, and more reliable. And then, almost before the gear could be used, the war ended.

Radio operators returned home. Before the war, many had been amateur radiotelegraphers, "hams" sending Morse code to other hams. Some had even managed to send voice or music. Now they had new ideas and experience with military triodes. Suddenly wireless voices sang out across America from stations with amateur or experimental call letters—8MK in Detroit with news, 9XM in Madison with weather, 6ADZ in Hollywood with music. Hams played and hams heard.

It was great, amateurish fun. But no major corporation—not Westinghouse, not the American Telephone and Telegraph Company, not even GE's newly created subsidiary, the Radio Corporation of America—had found any way to make money in broadcasting. None even marketed a broadcast radio receiver. Companies thought that money came from radiotelephony, from sending point-to-point messages.

Then a series of events in Pennsylvania changed everything. In his garage near Pittsburgh, Frank Conrad, who had designed mobile radios for Westinghouse during the war, put his old experimental station, 8XK, back on the air as soon as the government lifted its ban. Three weeks later, bored by talk, Conrad placed his phonograph in front of the 8XK microphone. Music played. Hams responded.

First, they requested favorite songs. Unable to answer each request, Conrad announced he would begin to broadcast. Next, the shop that sold him records offered to give them to him, in trade for free advertisements over the air. Then, late in 1920, a local department store ran an advertisement in the Pittsburgh *Sun* for a receiver that could pick up Conrad's wireless broadcasts—"on sale here $10.00 up."

At this point, in the brain of a vice president of the company that Conrad worked for—Westinghouse, of East Pittsburgh—a spark struck. Broadcasting made people want to buy wireless sets. Westinghouse could make wireless sets. Westinghouse must broadcast!

Within five weeks, thanks to Conrad, Westinghouse had built its first station in a rooftop shack in East Pittsburgh, won a broadcasting license and the commercial call letters KDKA, and entered the air with

Radio broadcasting took off in the 1920s. Since before World War I, radio buffs had been experimenting with homemade transmitters and receivers—many of them crystal receivers like the one operated by this Russian boy (left, upper). After the war, amateur broadcasts gained publicity, and radiotelegraph companies saw the profits in selling receivers and broadcasting. By 1930, 13.7 million *American homes owned radios.*

Novelty sold early models without much help from the salesman (opposite). The loudspeaker horn of this 1924 Radiola Superheterodyne replaced the crystal set's earphones. Sleeker styles like the 1934 Echophone (left, lower) developed as technology improved and demands changed. Some Prohibition-era radios even offered a hiding place for liquor.

196

"My friends," President Franklin D. Roosevelt would say warmly, and families like this one, hearing his "fireside chats," felt he was sitting with them in their living rooms. Radio created a political forum that Roosevelt exploited to win popular support for his New Deal programs in the 1930s. Radio networks also wooed listeners with comedy, music, sports, and drama. So popular was the cops-and-robbers serial "Gangbusters" (above) that its opening salvo of machine-gun fire and wailing sirens inspired a common phrase about making a knock 'em dead entrance— "coming on like gangbusters."

reports of the 1920 presidential election. Responses came back from as far as the Virginia shore. The press treated KDKA as big news.

Suddenly, at GE, RCA, AT&T, and across the country, the future of wireless became clear: commercial broadcasting. Stations went commercial almost overnight. Corporations built transmitting towers atop tall buildings. Westinghouse introduced its Aeriola receivers, and RCA its Radiola Superheterodyne, designed in part by inventor Edwin H. Armstrong. Businesses promoted themselves by hitching catchy slogans to their call letters: Sears, Roebuck and Co. opened WLS (world's largest store); the Chicago *Tribune* countered with WGN (world's greatest newspaper). Publishers and shopkeepers, teachers and preachers, labor organizers and mail-order quacks, all began to broadcast.

And the public responded. By 1924 a state college in Kansas had enrolled students from 39 states in its College of the Air. That year, 20 million people listened to national election returns transmitted from more than 400 stations. In 1927 the broadcast of a fight between Jack Dempsey and Gene Tunney allegedly caused ten fans to die of excitement.

Within a decade, radio had linked the nation. In 1922 only two-tenths of one percent of American homes had owned a radio receiver;

ten years later the figure was 60 percent of households, or more than 18 million radios. Recalling his own company's blindness before it saw radio explode in Pittsburgh, one researcher for GE seemed to speak for the entire country. "We had," he said, "everything except the idea."

Radio was destined, however, to do more than carry voices and music. The concept of radio detection and ranging, a process best known by its acronym, RADAR, came to scientists in many nations at much the same time, the 1930s. It began with two problems of very different sorts—a technical irritation and a national nightmare.

Over the years radio experimenters had hit the same technical problem: Radio waves bounced back when objects crossed their path. Some experimenters tried to turn the problem to advantage. A German in 1904 patented a radio obstacle detector. Two Americans in 1922 urged the Navy to guard its harbor approaches with radio burglar alarms. Such proposals, however, gained little support.

Then came the nightmare. Strategists in four nations—Britain, France, Germany, and the United States—realized they were heading for war. One nation had the nightmare most vividly. In the mid-1930s, in the face of war with Germany, Britain despaired at repelling the Luftwaffe's growing force of bombers. "The bomber," feared Prime Minister Stanley Baldwin, "will always get through." One man who disagreed was Henry Tizard, chairman of Britain's new Committee for the Scientific Study of Air Defence. His committee looked for ideas.

Many appeared. A professor of physics from Oxford suggested that trip wires, dangling from balloons or aircraft, be strung across the national air space. Assorted tinkerers turned out death-ray transmitters—blending such stuff as ultraviolet light and X rays into a beam that had allegedly knocked pigeons from the air. When one tinkerer asserted that powerful radio waves could knock down aircraft, Tizard's committee consulted the head of national radio research, Robert Watson-Watt.

The unconventional Watson-Watt did an analysis and proposed immediately that defenders of Britain prepare to aim radio pulses at hostile aircraft. The radio waves could do no damage, of course. But perhaps one-millionth of a millionth of a millionth of the force of the pulse would bounce back as an echo. By listening for these tiny echoes, Britain could know what was coming, from where, and how fast.

Britain's Air Ministry accepted Watson-Watt's proposal. In the spring of 1935, on a modest budget of ten thousand pounds, he dispatched four researchers to a semideserted military base on the North Sea. A month later, with a roomful of radar gear, they were tracking a plane 17 miles offshore. Within a year, on a budget of a million pounds, Watson-Watt's team began construction of five radar stations to guard London and moved their headquarters into a coastal mansion. At their radar estate, swims preceded lunch, cricket preceded dinner, and billiards in the evening competed with technical discussions before a roaring fire. Within two years, the researchers shrank their roomful of radar into a two-foot box, squeezed the box into a two-engine plane, and created airborne radar. When Britain entered the war against Hitler, British radar was more or less ready.

Radar helped win the Battle of Britain in World War II. As German fighters like this Messerschmitt escorted bombers over the white cliffs of Dover, radio pulses from aerial towers (opposite, upper) bounced off the planes and echoed back as blips on a screen (lower), revealing the enemy's range. Antiaircraft guns and British fighters then zeroed in. After 1940 a British-American research program developed the Microwave Early Warning system used in the D-day invasion of France.

Among myriad uses since then, radar tracks storms, guides ships, and maps another battle of the skies—air traffic around busy airports.

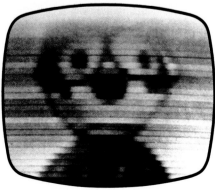

Felix the Cat auditions for television in 1928. Pirouetting on a phonograph turntable, the comic strip figure stars in experimental telecasts made by the National Broadcasting Company. A mechanical scanning device produced the blurred image at left. Electronic technology, refined over the next ten years, was needed to improve picture quality enough for a demonstration at the 1939 New York World's Fair.

Vladimir K. Zworykin (above) inspects his iconoscope, an electronic television camera tube, one of the inventions that brought him fame as the "father of television." Another television pioneer, Philo T. Farnsworth (above left, in glasses), demonstrates his 1935 electronic image dissector, part of a rival system.

OVERLEAF: A television studio audience watches a 1953 performance of "My Favorite Husband," while a technician adjusts the lighting. Audience response was part of the show, as it had been since radio days—comedians found empty studios unnerving.

After the war, Watson-Watt would repeat over and over the words of Adolf Galland, Germany's top ace in the Luftwaffe attack and later a general. The blame for the Luftwaffe defeat, said Galland, lay with Britain's "radar- and fighter-control network." In hindsight, it seems, the general saw it all: As German planes assembled over France, their "formations were already picked up and never allowed to escape from the [British] radar eye." British mastery of radar had hit the Luftwaffe as a total and "bitter" surprise. "We had," he said, "nothing like it."

In the number of years it took radar to move from idea to reality, the device it most resembled—another glowing tube that gave an instant but fuzzy picture of the world—evolved hardly at all. Television seemed to have the brightest future of all the devices that emerged from wireless. But year after year it kept failing to arrive.

Television's first nonappearance came in 1880. Alexander Graham Bell's telephone already allowed people to talk by telegraph. Early in 1880, Bell deposited at the Smithsonian Institution his secret plans for a new invention, the photophone. Rumors spread quickly: Bell had created the ultimate in instant communication—a way to *see* by telegraph.

Bits of the needed technology already existed. In the early 1870s, two British telegraph engineers had discovered a new property of the element selenium. Selenium conducted electricity more readily when hit with light than when left in darkness. The possibilities tantalized.

The human eye offered a model for electric vision. The eye detects not a unified picture, but rather myriad spots of light that combine like

An American couple shops for their first television set in 1950, joining the six million families who bought sets that year. World War II had halted manufacturing, but production surged in 1948, and with it network programming. Variety and talk shows, soap operas (originally radio serials sponsored by soap companies), news programs, and kiddie hours kept the customers buying.

Networks measured ratings mechanically to gauge programming. They found sports events had the greatest overall appeal. The crowd opposite, braving a downpour in 1951 to watch the first coast-to-coast telecast of a major-league baseball game, seems to bear this out.

Today such sports lovers would be watching or videotaping the game on their own sets. Some 98 percent of American households have at least one television set. Youngsters, says newscaster Edwin Newman, will have watched three years of television by the time they grow up.

tiles in a mosaic to suggest a complete image. By 1879 researchers in France, Ireland, and the United States had proposed mosaic systems for electric vision. George Carey of Boston, for example, devised an electrical circuit with selenium at one end and a light bulb at the other. If light hit the selenium, the bulb glowed. The next step would be to gather dozens of such circuits and create a mosaic at each end. If the shadow of someone's hand passed across the selenium mosaic, viewers of the light bulb mosaic would see a hand waving at them.

Such devices for turning light into electricity and back again (telectroscopes, some called them) amounted to only part of the pre-1880 ferment. Spinning discs and cylinders (fantascopes and praxinoscopes) made pictures seem to move. Sealed glass bulbs (cathode-ray tubes) glowed eerily when electricity flowed in them. Other spinning cylinders and wired microphones (phonographs and telephones, of course) carried human voices.

Anyone could imagine the next step. By 1879, predicted a cartoonist in *Punch,* all these devices would converge: The "telephonoscope" would allow Londoners to see all the way to farthest Asia. The world was surely ready for television.

Bell Labs

Somehow the words that launched the information age don't have quite the ring of "Eureka!" but still they deserve a place in history: "This thing's got current gain!"

That, an assistant recalled, is what Walter Brattain suddenly shouted in the lab they were sharing in December 1947 at that prestigious and unconventional research facility, Bell Telephone Laboratories, in Murray Hill, New Jersey. The "thing" Brattain had wired up was a slightly impure crystal of germanium, a type of semiconductor. It was multiplying the electrical current running through it—amplification, or current gain.

He realized at once what two years of work had yielded. He notified his colleagues John Bardeen and William Shockley. On December 23, 1947, the physicists demonstrated the first semiconductor device that could do the work of a vacuum tube: the transistor. It was to set off an avalanche of invention that would touch virtually everyone on Earth. It would also win the trio a Nobel Prize in Physics.

Not, however, the first Nobel at Bell Labs, nor the last. Since 1925, when American Telephone and Telegraph established its research subsidiary, Bell Labs has proved unusually fertile ground for discovery and invention, averaging one patent a day. Those inventions have transformed your daily life—probably your hourly life. Bell technology crops up in phone systems, of course, but also in movies, vitamin pills, television, computers, pressure-treated lumber, stereo, and on and on.

Bell Labs can do this partly by employing thousands of people in research and development. Most of them work in development, at several sites in New Jersey and elsewhere. Basic research—the sort of free-ranging investigation of the unknown that led to the transistor—takes place largely at Murray Hill. It exemplifies the real secret of Bell Labs' success—not just size, but a work environment unique in both industry and academia.

Unlike most corporations, Bell puts scientists, not business people, in charge of the administration of basic research. These managers in turn give subordinates free rein, invoking no deadlines or quotas—"managed anarchy," as one research chief puts it.

And, unlike most universities, Bell provides all the funding, so researchers need not waste time scrambling for grants. Furthermore, the workplaces themselves—floors and floors full of scientists—are a combustion chamber of ideas. In academia, a scientist might not learn of some critical discovery until a paper is presented at an annual meeting. Here it happens with casual ease. For physicist Harold Seidel the key to Bell's scientific success is the coffee urn. If you have a new idea, he says, "the best time to bounce it off somebody else is at the coffeepot."

One other thing has set Bell apart: patience. Pure research is an unpredictable road, with many hidden bends and forks. One discovery may not seem useful, but may lead to another that is. Kumar Patel, current research director, says many industrial labs overdirect. "They decide too early what they want to do" and so exclude other paths.

The winding road to the transistor and beyond illustrates his point. Clinton Davisson won Bell's first Nobel in 1937 for experiments he did in 1927 that revealed the nature of subatomic particles. His work proved that a bit of matter such as an electron can behave like a wave—a weird but fundamental principle of quantum mechanics that would help decades later in learning how semiconductors and lasers work.

Davisson hadn't planned on all that; he had simply been trying to figure out how electrons behaved in a vacuum tube. Vacuum tubes were essential to AT&T's growing long-distance telephone network. A phone signal weakened by journeying over hundreds of miles of lines could be amplified by vacuum tubes and sent on its way. But the tube had limitations. Something better was needed. Davisson's work had attracted a young physicist from MIT to Bell Labs—William Shockley. After World War II, Shockley headed up a group that included Brattain and Bardeen. Typically, their assignment

Genius at Bell Labs: inventors of the transistor John Bardeen, William Shockley, and Walter Brattain.

Worker assembles transistors in 1954.

was broad: Look into the little-known physics of semiconductors.

Their collaboration was characteristic of Bell Labs: Shockley supplied theoretical insight, Bardeen backed it up with mathematics, and Brattain put their ideas to the test. Not that the men didn't differ. Philip Foy, the assistant who overheard Brattain's exclamation that day, remembers Bardeen as the gentleman, always even tempered. But the contentious Shockley and irritable Brattain would often bicker. Foy recalled one of their progress meetings:

"Brattain was standing there clicking two quarters together. It was getting on Shockley's nerves. Brattain said, 'I can't think unless I'm clicking.' Shockley brings out two dollar bills and says, 'Here. Rub these together.'"

The transistor they devised, so much smaller and easier on power than vacuum tubes, was to spawn unimaginably smaller offspring. The semiconductor era would not only bring transistor radios to Bedouin caravans, but computers to outer space. It would render obsolete the very tools the team calculated with—their slide rules.

At first, such possibilities were hard to grasp. After the transistor's unveiling, *Popular Mechanics* daringly predicted a 20-fold shrinkage in hardware: "Computers in the future may have only 1,000 vacuum tubes and perhaps weigh only 1 1/2 tons." That computing power can now reside in a fingernail-size silicon chip, etched with tens of thousands of transistors.

Today, at the now huge Murray Hill building, labyrinthine hallways reflect frontiers of the 1980s and '90s. Tanks of liquid nitrogen stand here and there. Taped to lab doors are "Far Side" cartoons and signs warning of lasers in use. Some rooms glow under a ghastly yellow illumination that won't register on photosensitive materials. Somewhat ominously, shower heads and oxygen masks are spaced along the halls, in case something toxic should get spilled or inhaled. That's an improvement; few such precautions existed in transistor days when, by one report,

The Quiet Room provides a place to test sound equipment like these speakers, vintage 1953.

the lead drainpipes in the labs wore out every few months because researchers kept pouring acids into the sinks.

One room has not changed a lot since then: the anechoic chamber, or Quiet Room. Sound waves die here. Rather than echoing as in a normal room, they sink into the baffled walls, floor, and ceiling like a penny into snow. This sonically pure environment, so quiet you can hear your own pulse, provides a place to test sound systems now likely to be transistorized or computerized.

Many scientists feared Bell Labs would lose its treasured freedom after 1984, when AT&T had to give up its monopoly phone system and begin competing in the corporate marketplace. Changes did happen, but the flow of basic research papers from the Labs continues unstemmed. Evidently AT&T still believes that around every few bends on that winding road of discovery, something profitable may hide.

Transistorization made it easier, for instance, to loft electronic equipment into space, such as lunar landers and communications satellites—the latter of great interest to AT&T. Bell Labs designed the first really efficient communications satellite, Telstar, and the antennas that communicated with it.

It was while working with one of these horn-shaped antennas that Bell astrophysicists Arno Penzias and Robert Wilson discovered the echo of creation. They did not know it at the time; they just knew they were getting too much static. They fine-tuned everything. They found two pigeons living in the horn and evicted them. In 1965 they concluded the interference was real, a radio signal produced by heat. They had confirmed a controversial theory—that the universe began with an explosion. The heat they had detected was a residue of that Big Bang.

Thus the research labs set up simply to improve a phone system garnered another Nobel prize, for hearing the long-distance message of all time.

Jonathan B. Tourtellot

"Live via satellite" first flashed onto television screens after the launching of Telstar (test model at far right)— the communications satellite developed by AT&T's Bell Labs that first transmitted live programs overseas. Telstar, 34.5 inches in diameter and powered by solar cells, was boosted into orbit on July 10, 1962. That same day the satellite's double belt of antennas relayed telephone and television signals to and from ground stations on either side of the Atlantic.

"It all came down to the press of a red button in Maine," said project director Eugene F. O'Neill, referring to the 380-ton horn antenna (right) in Andover that transmitted and received Telstar's signals in the United States. So sensitive was its receiver that the horn could gather signals as weak as a billionth of a watt.

Satellite technology advanced rapidly beyond Telstar to produce a growing flock of orbiting "birds." They now carry most international television and global communications and have opened up other fields, such as cable television.

Finally, in mid-1880, Bell revealed his photophone. It used selenium. It worked with sound and light. But it did not transmit pictures; it used a light beam to transmit the human voice. The proper technology did not yet exist. Television would have to wait. And wait.

Television's development became a struggle of unflagging imagination against ever lagging technology. Two men in particular were experimenting with television at the same time on either side of the Atlantic. In 1923 in the United States, Charles Francis Jenkins made a public demonstration of a television system. Jenkins' television viewers saw the silhouette of a hand waving. Two years later in Britain, department store shoppers watched silhouettes produced on television by John Logie Baird. Both Jenkins' and Baird's systems used perforated discs based on the one invented by Paul Nipkow in 1884. But so long as television relied on mechanical devices such as spinning discs, it

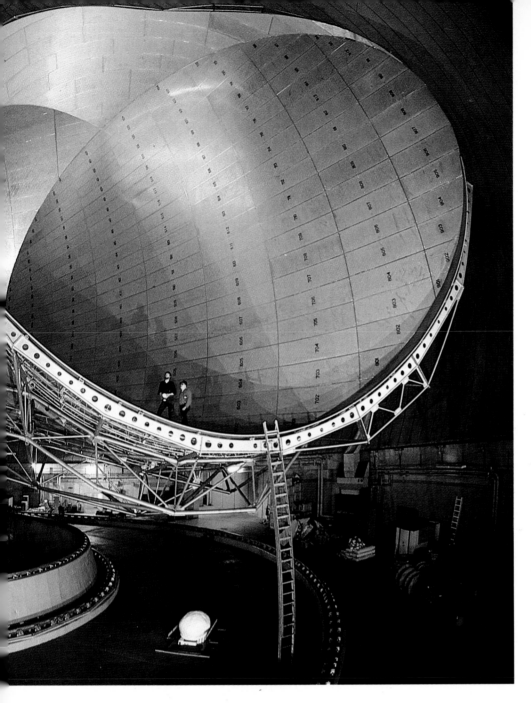

could not produce images of sufficient quality to satisfy the public.

The search for nonmechanical, all-electronic television evolved during the 1930s into a struggle in the United States between two very different men and their professional worlds. On one side was Vladimir K. Zworykin, a cosmopolitan and internationally educated engineer. As director of electronics research for RCA, Zworykin counted on corporate wealth and the support of RCA's visionary president, David Sarnoff. On the other side was Philo T. Farnsworth, a small-town and mostly self-trained inventor. He counted on modest funding from a few loyal supporters and on his own irrepressible drive.

Each man took his share of early steps in television's stumbling march. In 1923 Zworykin had made the first demonstration of an all-electronic television camera tube, using a mechanical transmitting device; it could barely transmit the letter X. In 1928 Farnsworth demonstrated the world's first all-electronic television system; it showed a man smoking a cigarette.

In 1933 Zworykin produced a major improvement on his 1923 design. The heart of his camera was a cathode-ray tube, resembling a hand lens on an angled shaft, which Zworykin called the iconoscope (from the Greek words for "image" and "to watch"). In it could be seen half a century of evolution from the telephonoscope fantasies of the 1870s. The iconoscope's light-sensing mosaic did not transmit its picture all at once to dozens of bulbs. Instead, a stream of electrons scanned rapidly across each of the mosaic's light-sensitive dots, creating a coded signal. Decoding, at the television's picture tube, then recreated an image in streams of glowing dots, each replaced so fast that the human eye saw a unified, moving picture.

Zworykin's iconoscope made high-quality television nearly complete. In Europe it dominated German broadcasting and formed the basis of British all-electronic broadcasting, inaugurated in 1936. But a problem remained for RCA. The firm needed some missing patents, without which Zworykin's iconoscope would lack sensitivity. Those patents belonged to Philo Farnsworth. The haggling took years. Finally, in 1939, RCA reached agreement with Farnsworth over a license on his innovations. Improved television broadcasting came almost immediately. So, too, came battles' over broadcasting standardization, and then came World War II and its ban on television manufacturing. But by that time, 60 years after Americans thrilled at the rumors of a photophone, television was ready to go, although not for years would American television service display the excellence already seen in Europe.

When the war ended, television's long-awaited entry into American life proceeded even more rapidly than radio's entry during the 1920s. In 1946 only two-hundredths of one percent of American homes had a television set. In ten years that number jumped to 72 percent, many of the sets tuned each week to "I Love Lucy" or "The $64,000 Question" or Edward R. Murrow's aptly named news show, "See It Now." In the 1960s television reached more than 90 percent of American homes and became the country's most trusted provider of information. Americans found television twice as credible as newspapers and five times more credible than radio. Watching, quite clearly, was believing.

And as Americans watched, sometimes before their eyes and sometimes behind the scenes, the wireless age continued its innovations. Television began broadcasting in color. Radio listeners discovered the rich sound of FM stereo. Communications satellites in orbit above the Equator linked all continents by wireless. Live television reached Earth from the surface of the moon. Radio transmitters on interplanetary spacecraft sent pictures of the outer planets back to Earth and then headed out of our solar system to become the first man-made objects in deep space. All the while, emitted from Earth's innumerable antennas, images of this planet, in sound and picture, preceded the spacecraft in their flight. For, as Marconi reminded his listeners long ago, wireless is not for our ears and our eyes only. "The messages wirelessed ten years ago," he said, "have not reached the nearest stars." But they will.

Spacecraft Voyager 1 trains its television cameras and radio antennas on the largest planet, Jupiter, in 1979. Launched by the National Aeronautics and Space Administration in 1977, Voyagers 1 and 2 have since been exploring the outer planets of our solar system, radioing back billions of pieces of data about new discoveries. By the end of the century, both spacecraft will have left the solar system. Scientists hope to continue tracking them for another 30 years.

Should they encounter intelligent life, the Voyagers carry copper discs with an electronic portrait of Earth: photographs of people, places, and animals; sounds of laughter, music, a rocket launch, a kiss; greetings in 55 languages. But communication with Earth will fade as they sail through outer space, directed there by radio— a technology only a century old.

Power Particles

By Richard Rhodes

O n a summer night in Paris in 1903, a group of friends who had gathered for dinner retired to their host's garden to enjoy the late evening air. One member of the group, the Polish-born physicist Marie Curie, had been awarded her doctorate of science that day. Five years before, she and her husband Pierre had discovered two new elements, radium and polonium, discoveries for which they would soon receive the Nobel Prize in Physics. They had come to the dinner party prepared to show off one of the elements.

When the group was settled in the garden, Pierre brought out a liquid-filled tube glowing with light. The tube contained a strong radium solution and was coated with zinc sulfide, which the radium caused to fluoresce. "The luminosity was brilliant in the darkness," another guest, the British physicist Ernest Rutherford, remembered, "and it was a splendid finale to an unforgettable day." Pierre Curie's hands, Rutherford noticed ominously, "were in a very inflamed and painful state due to exposure to radium rays." That early, nuclear energy revealed both its promise and its potential for destruction.

The two elements the Curies had identified were radioactive (a word Marie herself coined). Radioactivity had been the discovery of French physicist Antoine-Henri Becquerel, an expert on fluorescence. Early in 1896 Becquerel had heard a report on the work of Wilhelm Röntgen, the discoverer of X rays, and had decided to search for materials that might release such rays as the glass walls of Röntgen's X-ray tubes did. One material Becquerel tried was a uranium salt. Sprinkling it on a sealed photographic plate, Marie Curie reported, "he obtained photographic impressions through black paper."

When the penetrating rays that poured forth from such seemingly inert matter were sorted out, they proved to be not X rays but three distinct types of radiation. Unlike X rays, these alpha, beta, and gamma rays needed no external bombardment to release them. Ernest Rutherford calculated the energy of their outpouring from a given lump of radium to be a million times greater than the energy released from the same amount of material if it were burned. What was the source of such prodigious energy, awed physicists wondered, and how might it be controlled? Answering those questions was the work of the next four decades, work that culminated in the operation of the first nuclear reactor and the explosion of the first atomic bomb.

Ernest Rutherford's research was crucial. He had come to Cambridge University on a scholarship from New Zealand in 1895, just one month before Röntgen's world-celebrated discovery, and he quickly made the new field of radioactivity his own. He was a big, vigorous man, bold, with a loud voice, who his students would remember

In 1952 a hydrogen bomb, the first of its kind, exploded on a Pacific atoll with energy equivalent to ten million tons of TNT. Just five decades earlier, Albert Einstein and other physicists had begun theorizing about the power locked inside the nucleus of an atom. The pressures of war swiftly propelled their ideas out of the laboratory with the invention in 1942 of a means to harness this power: the first nuclear reactor. The atomic bomb soon followed, destroying Hiroshima and Nagasaki in August 1945 and introducing a new kind of war.

"It did not take atomic weapons to make war terrible," said J. Robert Oppenheimer, the brilliant physicist who directed the building of the first atomic bombs. "But the atomic bomb was the turn of the screw. It has made the prospect of . . . war unendurable."

Potent in peace as well as war, atomic energy generates power for electricity, propels ships and submarines, and creates products for use in medicine, agriculture, and industry.

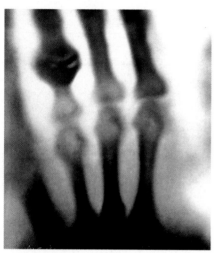

Insides revealed! German physicist Wilhelm Röntgen in turn-of-the-century caricature displays evidence of his discovery (upper): X rays, a form of radiation able to penetrate flesh and now known to arise from changes in an atom's nucleus. Röntgen made his first X-ray photograph in 1895, an image of his wife's hand (lower). A year later, Antoine-Henri Becquerel found that the element uranium emitted radiation spontaneously—a phenomenon called radioactivity by Marie Curie (opposite) and her husband, Pierre. The Curies' work with radium greatly enhanced understanding of radiation as a source of power.

marching around the laboratory singing "Onward Christian Soldiers" to fire their enthusiasm. It was he who first identified and named alpha and beta radiation.

To study the structure of the atom, physicists bombard matter with particles and measure how the particles react. Rutherford used alpha particles from radium for his experiments. At the turn of the century, no one knew what an atom looked like. Rutherford's teacher, J. J. Thomson, had proposed for an atomic model a diffuse, positively-charged sphere stuck with negative electrons—the "plum-pudding model," it came to be called. In theory, positively-charged alpha particles from radium were energetic enough to pass through such a structure, only slightly deflected by the repulsion of its positive charge, and usually they did.

But experiments bombarding thin gold foils that Rutherford was conducting in 1909 seemed to be troubled with stray particles. Fixing up the equipment didn't make the strays disappear. Rutherford sensed opportunity. He asked one of his assistants, Ernest Marsden, to look for particles bounced back from the foil. Marsden found the recoils of Rutherford's hunch. "It was almost as incredible," the master said later, "as if you fired a 15-inch [cannon] shell at a piece of tissue paper and it came back and hit you." The atom, Rutherford announced, was not a diffuse cloud of positive electricity stuck with electrons. It had a minute core, a nucleus, so dense that it deflected the radium alphas that directly approached it around and back the way they came, as the sun swings an approaching comet around again and out. All the vast energy of radioactive decay came from that tiny nucleus, a fly in a cathedral compared to the shells of orbiting electrons that surrounded it.

Bombarding the nucleus for study was difficult. Not many alpha particles could push through the positively-charged electrical barrier that surrounded it. What physics needed was a new particle that carried no electrical charge. The two known building blocks of matter were electrons and protons. Protons seemed to make up the nucleus but failed to account for all of its mass in heavier elements like uranium. Rutherford speculated in 1920 that a third basic particle might exist, a neutron, as heavy as the proton but electrically neutral.

A Nobel laureate by this time, Rutherford had assumed the directorship of Cambridge University's Cavendish Laboratory. His protégé James Chadwick—a tall, dark, wiry man with a raven's-beak nose—came along as assistant director. Intrigued by Rutherford's speculation, Chadwick spent his scarce spare time over the next decade searching for the neutron.

A mistake led to breakthrough. The Curies' daughter Irène had become a physicist. Working in Paris with her husband, Frédéric Joliot, she was trying to identify a strange, highly penetrating radiation that the light metallic element beryllium released under alpha bombardment. The Joliot-Curies thought it was gamma radiation, a form of light more penetrating than X rays. Early in 1932 they reported that it knocked high-velocity protons out of paraffin wax.

Chadwick was skeptical of their report. Gamma rays could certainly

Ernest O. Lawrence

When Ernest Orlando Lawrence, age 18, arrived home from college for the summer of 1920, he purchased an old Ford for $125. By August he owned a brand new Ford and was not a penny poorer: He had bought a string of used cars, repaired them, and sold each for a profit. By September he had sold the new Ford at a profit, too.

Eleven years later, at the University of California at Berkeley, Lawrence split a lithium atom to create helium with a machine he invented and called a cyclotron. In the following decade, Lawrence built a string of cyclotrons, each more powerful than the last, and used them to smash a variety of atoms. He created radioactive isotopes that he and his brother John used to battle their mother's cancer. He discovered radioisotopes of common elements.

The cyclotrons worked on the principle of electromagnetic resonance: The magnetic field generated by a powerful electromagnet guided charged particles in a circular path within a vacuum chamber. With each revolution, two electrodes boosted the velocity of the particles, spiraling them from the center of the chamber to the edge, where they finally shot out to smash their atomic target. By 1941 he had separated fissionable Uranium 235 from Uranium 238 and created a prototype for the electromagnetic separation machines at Oak Ridge, Tennessee, that helped produce enough Uranium 235 for the bomb dropped on Hiroshima.

"There's *nothing* you can't do if you work hard enough," Lawrence once said to his brother. Lawrence at work was like the charged particles spinning in his cyclotrons—everyone he worked with felt his force. "He pulled others into his orbit," said Warren Weaver of the Rockefeller Foundation. Ernest Lawrence never stopped moving.

Lawrence at cyclotron controls in 1938.

He couldn't stop moving in the hours before dawn on July 16, 1945, just before the first atomic bomb was tested in New Mexico. As the assembled scientists waited for the fruits of their labor to sear the desert sky, Lawrence climbed in and out of a car, unable to decide where he wanted to be at the moment of truth. When the truth was known, and some of his companions stood paralyzed with fear for the future of humanity, Lawrence hopped backslapping from one to the next, shouting "It works! It works!"

He had little time for the political and moral dilemmas he labeled "causes and concerns." Likewise, he rarely spent time on the theory of physics. When he found his researchers studying equations at a blackboard, he ordered them back to work.

They couldn't fault him though—Lawrence did so much work himself. Not the least of it was getting the money he needed to build bigger cyclotrons. He was a charming, persuasive fund-raiser. And a committed one. When he won the 1939 Nobel Prize in Physics for the invention of the cyclotron, he arranged to receive the award in Berkeley rather than in Stockholm to avoid losing a month in his fund-raising campaign to finance the next cyclotron. On February 29, 1940, he accepted the prize, and ended the ceremonies with a plea for money.

Ernest Lawrence set goals for himself and then did whatever he had to do to achieve them. His first efforts focused on electrical apparatus. When he was a schoolboy in Pierre, South Dakota, he drilled holes through the dining room table, installed wires and a telegraph key, then ran wires to Bernard Murphy's house next door so the boys could practice code. When he entered the University of South Dakota in 1919, his head full of wireless communication, he went straight to Lewis Akeley, dean of the College of Electrical Engineering, to persuade him to buy wireless equipment for the school.

Akeley was more interested in Ernest Lawrence than in a wireless communication system. He wondered why a boy so eager to experiment was not enrolled in physics courses. Unwilling to interfere, he waited patiently for Lawrence to find the physics department, but Lawrence never did. All he could think about was wireless communication. Finally, at the end of the school year, Akeley said to Lawrence, "If you come back here and spend the month of August with me, before school opens, I'll interest you in physics. If I can't, I'll never speak of it to you again." Lawrence came back, fell in love with physics, and that was that.

Ernest Lawrence always pressed forward, never looked back. Maybe it was his South Dakota roots, his midwestern temperament, his forthright personality. A personality that made him always see things in black and white. Even college love letters to his girlfriend, Ruby Patterson, left no doubt about his moralistic positions on such issues as wasting time, worthy goals, jealousy, fidelity, and promiscuous necking. The sides of an issue, all two of them, were always clear to Ernest Lawrence. And he always knew which side he was on.

When the stakes were much higher—when it was a matter of life and death—Lawrence was the same way. Of the scientists involved in the Manhattan Project, he was among those who held out longest against the military use of the bomb against Japan. He felt that a planned demonstration of its force would be enough to end the war . . . all wars. When he was finally convinced otherwise, he never turned back. After the horror of Hiroshima, when other scientists questioned the value of creating and testing new, more powerful nuclear weapons, Lawrence supported U. S. development and stockpiling of nuclear armaments.

True to character, Ernest Lawrence worked himself to death, insisting on staying in Geneva at a 1958 conference until he was near collapse with ulcerative colitis. He had to be rushed home to Berkeley. The man who never stopped working finally did. But the

Lawrence's notes and 1931 cyclotron.

way he worked all his life—the passion he devoted to his experiments and machines—is what he is remembered for. A colleague once described the Lawrence modus operandi: "The system was something like this: . . . while we were doing research with the 27-inch [cyclotron] and the 37-inch was being designed, he was dreaming up the 60. While we were using the 37-inch and the 60-inch cyclotron was being designed the 184 was being dreamed of by Ernest Lawrence."

Lynn Addison Yorke

Lawrence, at his Berkeley laboratory in 1934, adjusts the 27-inch cyclotron.

A lab blackboard announces the winner of the 1939 Nobel Prize in Physics.

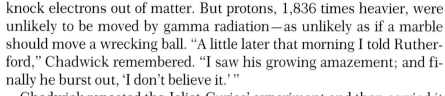

knock electrons out of matter. But protons, 1,836 times heavier, were unlikely to be moved by gamma radiation—as unlikely as if a marble should move a wrecking ball. "A little later that morning I told Rutherford," Chadwick remembered. "I saw his growing amazement; and finally he burst out, 'I don't believe it.'"

Chadwick repeated the Joliot-Curies' experiment and then carried it a step further. The strange radiation was certainly powerful. Chadwick used it to bombard nearly a dozen elements, and it knocked protons out of every one. The protons ejected with far more energy than gamma rays could muster. But a particle with about the same mass as the proton could do the job, Chadwick saw. It would have to be electrically neutral to penetrate materials as effectively as it did. "We may suppose it [to be] the 'neutron' discussed by Rutherford," Chadwick wrote in triumph. Chasing down a new elementary particle, work that would earn him a Nobel Prize in Physics, he had hardly slept in ten days. He stumbled to a meeting of his Cambridge colleagues to report his great discovery. "Now I want to be chloroformed," he finished his brief talk abruptly, "and put to bed for a fortnight."

The neutron could penetrate matter far more effectively than any other known particle, but the leading physicists of the day—Rutherford, Albert Einstein, and Niels Bohr—discounted the possibility that any conceivable mechanism could serve to release useful quantities of energy from the atom.

One of Einstein's young protégés disagreed. Leo Szilard, a brilliant Hungarian physicist, had left Nazi Germany for London in 1933. One gray September morning he read in the London *Times* that Rutherford had denounced talk of atomic power as "moonshine." Szilard went walking to think the problem through. Crossing the street, he realized suddenly that an element might be found "which is split by neutrons and which would emit *two* neutrons when it absorbs *one* neutron." Such a process might initiate a runaway geometric progression—a chain reaction, Szilard named it, borrowing the term from chemistry— one neutron releasing two, two releasing four, four releasing eight, the numbers doubling with each generation, and energy coming off with each reaction as well. "The idea never left me," he said later. "In certain circumstances it might be possible to set up a nuclear chain reaction, liberate energy on an industrial scale, and construct atomic bombs."

But which element among the 92 known in nature, from hydrogen to uranium, might chain-react? Szilard didn't know and wasn't prepared

Particles accelerated to high velocity by a cyclotron at the Argonne National Laboratory in Illinois generate a blue glow. American physicist Ernest O. Lawrence invented the atom-smashing cyclotron in the early 1930s to create high-speed particles that penetrate atoms and unleash their energy. Today the Argonne cyclotron makes radioactive isotopes for use in medicine and research.

Neutrons are impossible to see, but their effects are not. In 1932 Frédéric Joliot-Curie (opposite, at right) produced the first evidence of these electrically neutral particles in a photograph like the one above, which shows the spidery tracks of oxygen and carbon set in motion by neutrons. A few years later, two European scientists, Otto Hahn and Lise Meitner (above right, at center), decided to see what would happen if they bombarded the radioactive element uranium with neutrons. Meitner was forced to flee the Nazis in 1938, but Hahn continued their work and in a letter to Meitner late that year described some astonishing results. Meitner guessed their significance: The neutrons had split the uranium nucleus, releasing 200 million electron volts of energy.

Later, scientists found that this process, fission, also spawned additional neutrons that might split neighboring nuclei, creating a chain reaction—an avalanche of fissions accompanied by the release of huge quantities of energy.

to find out. He was a refugee, living off his savings, without access to a research laboratory. Rutherford heard his appeal unmoved. Szilard impressed Winston Churchill's science adviser, Oxford's Frederick Lindemann, sufficiently to win an offer of employment. But no one in England was interested in nuclear chain reactions.

That worried Szilard. Scientific knowledge, he knew, was open knowledge; what he had realized, some other physicist might realize as well—and what if that physicist was a German, with all the resources of the Third Reich at his disposal? The specter of the Third Reich made invulnerable with atomic bombs filled Szilard with concern. He patented his idea and gave the patent to the British Admiralty to protect as an official secret; the Admiralty accepted the gift and filed it away.

War was approaching. In Germany and elsewhere, scientists probed the atom with the neutron. Neutron bombardment could transmute one element into another. It could also make variant physical forms of the same element—isotopes. At Berlin's Kaiser-Wilhelm Institute for Chemistry in the late 1930s, the German chemist Otto Hahn and the Austrian physicist Lise Meitner used neutrons to bombard the elements and sorted out the resulting transmutations chemically.

Hahn and Meitner found no element more puzzling than uranium. In its natural form uranium was predominantly a mixture of two isotopes—one part rare U235, 139 parts U238. But bombarding natural uranium with neutrons produced strange and inconsistent mixtures of different substances. The two researchers suspected they were producing new, man-made elements heavier than uranium, materials never seen before. That would be a discovery of deep originality; they labored long and hard to prove it. But the more substances they traced chemically, the more confusing the physical explanation became.

Although Meitner had been baptized a Protestant, by Nazi law she

226

was a Jew. With the annexation of Austria to the Third Reich in March 1938, her future in Germany looked bleak. Hahn was no Nazi. He and Dutch friends helped the shy, 60-year-old woman escape to Holland that summer and then to Sweden. There she languished unhappily while Hahn and the young chemist Fritz Strassmann continued the uranium work in Berlin.

At Christmastime 1938 the breakthrough came. Hahn and Strassmann bombarded a solution of uranium nitrate with low-energy neutrons and discovered they had made barium, an element little more than half as heavy as uranium. Since the energy of the bombarding particle usually determined how much matter it would chip off the nucleus it struck, that outcome was unexpected and astonishing. Hahn wrote Meitner the incredible news.

She was vacationing then near the west coast of Sweden with her physicist nephew Otto Frisch. Together the two physicists worked out an explanation. It was a case of the straw that broke the camel's back: The uranium nucleus, unstable to begin with, had absorbed the small added energy of the bombarding neutron and split in two, its fragments reforming as barium and krypton. Hiking in the holiday snow, aunt and nephew worked the numbers and realized that each bursting uranium atom would release nearly 200 million electron volts of energy— enough energy from each atom, Frisch estimated later, to make a grain of sand jump. Fission, the physicists named the reaction, borrowing the term from biology.

The physics world was stunned. Leo Szilard, among others, immediately saw the possibility of the explosive chain reaction he had imagined back in 1933. At Columbia University he performed a simple experiment bombarding uranium oxide and found roughly two neutrons released for each one neutron absorbed. "That night," Szilard remembered later, "there was very little doubt in my mind that the world was headed for grief."

The Second World War began in Europe with the Nazi invasion of Poland on September 1, 1939. Since nuclear fission was no secret, and it was evident to physicists that a weapon that harnessed such enormous power would be decisive militarily, every major industrial nation moved to sponsor research toward an atomic bomb. As it turned out, only the United States had the industrial capacity to conduct such a monumental enterprise in the midst of total war. What began as a few scribbled formulas on a blackboard in 1939 became, by 1945, an industrial plant that rivaled the U. S. automobile industry in scale.

The Italian physicist Enrico Fermi, a friendly man with a wry sense of humor who had emigrated to the United States to protect his Jewish wife from Italian anti-Semitism, contributed decisively to the top secret Manhattan Project. He advised on the design of the first atomic bombs, but his more crucial contribution was performing the historic physics experiment demonstrating that a chain reaction might be made self-sustaining and also controlled.

With Szilard locating supplies of graphite and uranium, Fermi and a team at Columbia University began planning what Fermi called a pile,

Nazi faithfuls attend a nighttime rally at Nürnberg in 1938. The specter of Nazi power bolstered by nuclear weapons worried many American and European scientists. In 1939 Albert Einstein signed a letter to President Roosevelt: "Sir: Some recent work . . . leads me to expect that the element uranium may be turned into a new and important source of energy in the immediate future . . . and it is conceivable . . . that extremely powerful bombs of a new type may thus be constructed." Roosevelt promptly appointed a committee to investigate the possibility of building an atomic bomb. The nuclear race was on.

Enrico Fermi

In a predawn chill amid the scrublands of Alamogordo, New Mexico, a short, muscular man lay belly-down in a shallow ditch, waiting. July 16, 1945. 5:29 a.m. Suddenly, blinding white light filled the sky. Enrico Fermi stood up, clutching a fistful of paper bits, and turned to watch a huge fireball nine miles away swirl and boil upward.

Seconds passed. Fermi opened his fist to let some of the paper bits drift to his feet. Then a powerful blast of air hit, pressing hard against Fermi's clothing and skin. He let more paper fall. It skittered away from the blast, landing on the ground at a distance. Fermi measured the distance and made a quick calculation. By this crude method, the Italian physicist instantly estimated the energy liberated by the world's first nuclear explosion.

A passion for quantifying, for going straight to the facts and arriving there first: These were the hallmarks of Enrico Fermi's genius. Born in Rome in 1901, Fermi discovered his love of mathematics and physics as a child. Propelling himself rapidly through secondary school, he entered a special university physics program at the age of 17. There he wrote to a friend, "In the physics department I am slowly becoming the most influential authority." So much an authority, in fact, that the department director begged Fermi to instruct him in theoretical physics. "I am an ass," the director told him, "but you are a lucid thinker and I can always understand what you explain."

Fermi strove for clarity, constantly measuring and classifying things to enhance his understanding. "[His] thumb was his always ready yardstick," wrote his wife and biographer, Laura Fermi. "By placing it near his left eye and closing his right, he would measure the distance of a range of

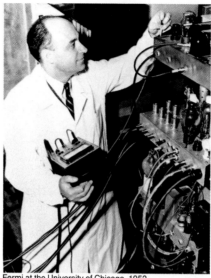
Fermi at the University of Chicago, 1952.

mountains, the height of a tree, even the speed at which a bird was flying."

A brilliant theorist, Fermi discovered a major principle of gas behavior and, in a definitive work on beta radiation, explained one of nature's four basic forces—all before the age of 33. At the same time, he indulged his passion for concrete problems.

Just such a problem presented itself in 1934, when Fermi—then a tenured professor at the University of Rome—learned that scientists had found a way to create radioactive elements by bombarding nonradioactive ones with positively-charged alpha particles. Fermi had a hunch that neutrons, which had no charge, would produce radioactivity more effectively. To prove it, he and his team filled small hollow cylinders with elements from hydrogen to uranium, and placed them in a lead housing. Then they bombarded the elements with neutrons at one end of the university physics hall and measured their radioactivity with a homemade Geiger counter at the other end. Clasping the short-lived radioactive products, Fermi raced his colleagues down the hall, lab coat flying—and took childish delight in declaring himself the winner.

In a moment of inspiration, Fermi took the experiment a step further: He

removed the lead separating the neutron source from the elements and replaced it with paraffin. The Geiger counter went wild, and the physics hall erupted in cheers: "Fantastic! Incredible! Black magic!" The light paraffin had slowed the neutrons, and—as Fermi theorized later—slow neutrons enter a nucleus and produce a reaction more easily than fast ones, just as a slow-rolling golf ball is more likely to drop into a hole than a fast one. Though he didn't realize it at the time, Fermi had discovered a key to harnessing nuclear energy. The work won him the 1938 Nobel Prize in Physics.

That year the physicist, his Jewish wife, and their two children left Fascist Italy for the United States. Shortly after their arrival, some stunning news reached Fermi: Two German scientists had shown that the bombardment of uranium with neutrons actually fissions, or splits apart, the uranium nucleus, releasing energy and spawning more neutrons. Fermi realized immediately that, given the right conditions, uranium fission might give rise to a self-perpetuating nuclear chain reaction—and an accompanying release of enormous quantities of energy. He stood at his office overlooking Manhattan. "A little bomb like that," he said to himself, cupping his hands as if holding a ball, "and it would all disappear."

Fermi went to work, first in New York, then at the University of Chicago, to assemble a device that would make possible a chain reaction. The reactor, or pile, slowly took shape: a huge, layered lattice of grimy black graphite bricks impregnated with lumps of uranium. On December 2, 1942, Fermi and his colleagues conducted the final stage of the experiment in a squash court beneath the stands of an abandoned football field in Chicago's South Side. Standing on a balcony adjacent to the ceiling-high pile, Fermi ordered a control mechanism—a rod of neutron-absorbing material—withdrawn from the pile by a few inches, then by a few more and a few more. At 3:49 p.m. the pile "went

In a University of Chicago squash court, Fermi and his colleagues oversee the operation of the first chain-reacting pile, December 2, 1942.

critical": no fireworks, no big bang, just the silent motion of a neutron recorder as the neutrons multiplied, cleaving more and more uranium atoms. The man who loved to be first had created the world's first self-sustaining nuclear chain reaction. The race to build an atomic bomb was on.

With the help of Fermi's calculations, the bomb went rapidly from drawing board to production. In July 1945 an atomic bomb exploded on that remote desert in Alamogordo, New Mexico, and a month later, two more destroyed the Japanese cities of Hiroshima and Nagasaki.

After the war, Fermi returned to Chicago to teach and to continue his work on the atomic nucleus. His health began to fail in the summer of 1954. With characteristic willpower, he tried to sustain his usual pitch of activity, lecturing, teaching, hiking. But in October he learned that he had terminal cancer. His friend and biographer, Emilio Segrè, visited Fermi in the hospital shortly before his death. "In typical fashion," Segrè wrote, "he was measuring the flux of nutrients [from a feeding tube] by counting drops and timing them with his stopwatch." Fermi died in November 1954.

Jennifer G. Ackerman

Fermi's lab at the Rome Institute of Physics, 1934.

July 1945: In the remote desert north of Alamogordo, New Mexico, a crew prepares to hoist an atomic bomb to the top of a steel tower, raising it 100 feet in the air to reduce the amount of sand sucked up when the bomb explodes—sand that would later rain down to earth as radioactive debris. Made of plutonium, the bomb produced the first man-made nuclear explosion, code-named Trinity.

the first nuclear reactor. After the Japanese attack on Pearl Harbor on December 7, 1941, and the entry of the United States into the war, they moved to the University of Chicago and prepared a definitive experiment, Chicago Pile Number 1, to be assembled in an abandoned squash court under the stands of the university football stadium.

Fermi's determined team had to stack together, by hand, 771,000 pounds of graphite bricks, 80,590 pounds of uranium oxide pressed into stubby cylinders and grapefruit-size spheres, and 12,400 pounds of uranium metal cast as cylinders. The pile went up in layers, one layer of solid graphite followed by one layer of graphite bricks drilled with holes to house the uranium. Uranium produces a few neutrons spontaneously; the graphite—pure crystalline carbon—gave the neutrons something to bounce against to slow them down. Properly slowed, they encountered atoms of the rare isotope U235 in the natural uranium and fissioned them, releasing more neutrons and furthering the chain reaction. Slots in the graphite bricks allowed neutron-absorbing cadmium rods to be moved in and out of the pile, controlling its rate of reaction. When the pile was large enough, more neutrons would be released within its bulk than were lost at its surface, and the reaction would be self-sustaining. Two crews worked around the clock to build the strange machine. Graphite dust blackened the walls and floors; the crews stumbling home to sleep looked like coal miners. Winter locked down and the squash court turned bitterly cold. Finally, on the night of December 1, 1942, the work was done.

December 2 dawned cold and gray. A harsh wind blew. Observers gathered in their overcoats on the squash-court balcony to watch. Fermi ran measurements, ordered the control rod removed inches at a time, checked his calculations. He stopped the experiment for lunch and began again at two in the afternoon. Before lunch the control rod had been seven feet out. Fermi ordered it out a foot more. "This is going to do it," he said. "Now it will become self-sustaining." An eyewitness remembers "the sound of the neutron counter, clickety-clack, clickety-clack. Then the clicks came more and more rapidly, and after a while, they began to merge into a roar." The Italian laureate turned to watch the increasing reaction on a silent chart recorder. "Suddenly Fermi raised his hand," the eyewitness continues. "'The pile has gone critical,' he announced."

Another physicist in attendance that historic day, Hungarian émigré Eugene Wigner, remembers the team's reaction: "For some time we had known that we were about to unlock a giant; still, we could not escape an eerie feeling when we knew we had actually done it. We felt as, I presume, everyone feels who has done something that he knows will have very far-reaching consequences which he cannot foresee."

Fermi's experiment had proved that a slow-neutron chain reaction could be made self-sustaining in natural uranium. But could a fast-neutron bomb of pure U235 be built? Could the new man-made element plutonium, bred from uranium in industrial-scale versions of Fermi's pile, be made into a bomb as well? The challenge of answering those questions fell to a tall, slim, chain-smoking American physicist

.006 seconds

.025 seconds

.100 seconds

15 seconds

"I am become Death, the shatterer of worlds." The sacred Hindu words came to Los Alamos Director J. Robert Oppenheimer as he watched the explosion of the world's first atomic bomb. On July 16, 1945, the ball of plutonium at the heart of the bomb was violently squeezed by an explosive, instantly unlocking the nuclear force holding together billions of atoms. First came a burst of light, "a great green supersun," wrote the only reporter allowed at the top-secret site, "climbing . . . to a height of more than 8,000 feet," then spreading out in a vast mushroom cloud. The blast was estimated at 18,600 tons of TNT.

Months later, Oppenheimer (above, at left) visited the site with Leslie R. Groves, the U. S. Army general in charge of Trinity, and found only the twisted remains of the tower footings.

named J. Robert Oppenheimer, a wealthy Manhattanite who taught at the University of California at Berkeley. The Army chose Oppenheimer to direct the top secret laboratory where the atomic bombs would be designed and assembled. Oppenheimer sited the new laboratory, Los Alamos, in a forested wilderness northwest of Santa Fe, New Mexico.

By 1945 Oppenheimer guided a staff of several thousand men and women at Los Alamos in the development of two kinds of bombs: one for uranium, one for plutonium. The uranium design was relatively simple, but inventing the plutonium system, a physicist who worked on it believes, was "the most difficult technological feat ever attempted up to that time." In Tennessee vast separation plants extracted rare U235 from natural uranium. Ernest O. Lawrence, inventor of the atom-smashing cyclotron, directed the development of one important extraction system. Though the Tennessee plants worked almost nonstop for years, the processes were so inefficient that the uranium bomb, Little Boy, would be one of a kind and would go into the war untested—to be exploded over Hiroshima on August 6, 1945. The plutonium bomb needed smaller quantities of fissionable material. In the winter of that final year of war, shipments of plutonium began arriving at Los Alamos from the big production reactors built beside the Columbia River at Hanford, Washington. Fat Man, the type of bomb that would be exploded over Nagasaki, was complicated enough to require a test.

Los Alamos prepared Trinity Site, a testing area in the barren desert north of Alamogordo, New Mexico. The world's first atomic bomb

Researchers at Los Alamos, New Mexico, manipulate radioactive isotopes (above). One of the first peaceful applications of atomic energy, these radioactive by-products of fission are produced in cyclotrons and nuclear reactors for use in medicine, agriculture, and industry. An isotope's radioactivity acts as a "tag" that permits scientists to trace its path through chemical and physical changes—in the body, for instance, or in crops.

Another military use for nuclear energy appeared in the mid-1950s: propulsion for submarines and ships. The nuclear-powered submarine U.S.S. William H. Bates (opposite) can cruise the ocean for months without refueling.

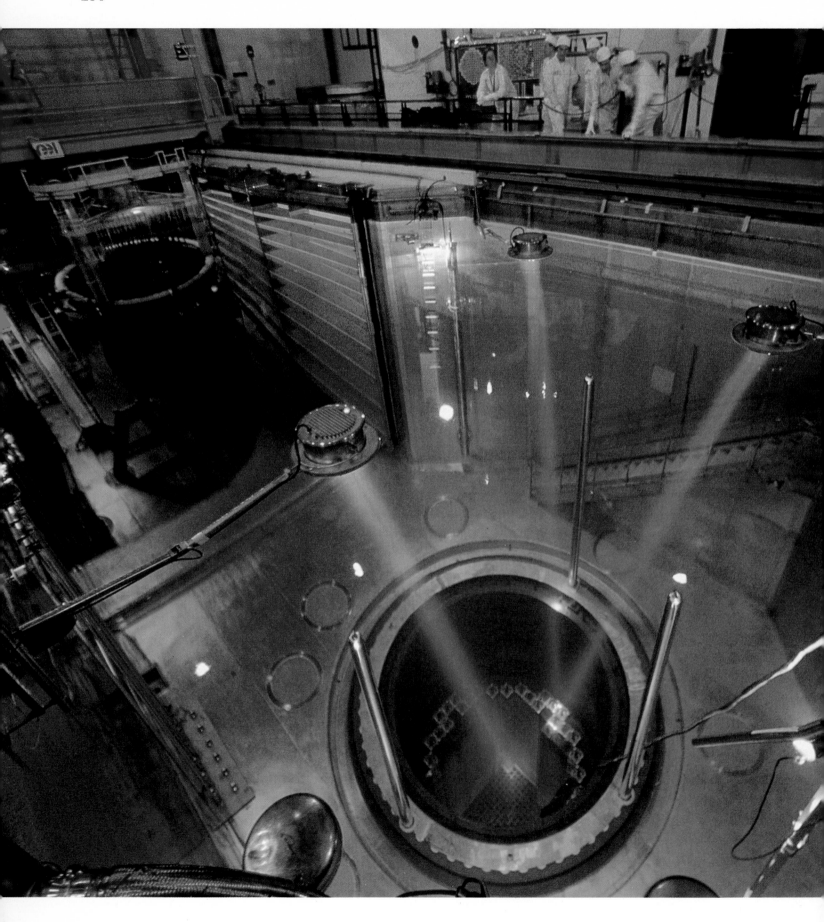

235

arrived partly assembled on Friday, July 13. By Sunday the two-ton weapon had been hoisted to the top of a hundred-foot steel tower.

Thunderstorms Sunday night delayed the test. The shot was finally fired just before dawn on July 16, 1945, while observers waited in bunkers five miles away or lay pressed against the ground in trenches ten miles away. Physicist I. I. Rabi described the world's first nuclear explosion, equivalent in explosive force to 18,600 tons of TNT:

Suddenly, there was an enormous flash of light, the brightest . . . I have ever seen. It blasted; it pounced; it bored its way right through you. It was a vision which was seen with more than the eye. It was seen to last forever. You would wish it to stop; altogether it lasted about two seconds. Finally it was over, diminishing, and we looked toward the place where the bomb had been; there was an enormous ball of fire which grew and grew and it rolled as it grew; it went up into the air, in yellow flashes and into scarlet and green. It looked menacing. It seemed to come toward one.

A new thing had just been born; a new control; a new understanding of man, which man had acquired over nature.

A month later Little Boy and Fat Man destroyed two Japanese cities and ended a long and terrible war. The development of the vastly more powerful hydrogen bomb, first tested by the United States in 1952 and by the Soviet Union late in 1955, led to an unprecedented arms race but also to a remarkably stable condition of armed truce.

The first peaceful uses for nuclear energy had appeared in the 1920s: radioactive isotopes for medicine and industry. Other peaceful applications emerged slowly after the war, delayed in the United States by the necessity of developing legislation to control the new technology and by the mistaken conviction that uranium ore existed in usable form in only limited quantities and should be reserved for weapons production. Civilian nuclear power came in the mid-1950s as a by-product of another military application developed after the war: power for a new class of submarines. A diminutive, hard-charging U. S. Navy engineer named Hyman Rickover pushed both programs through.

The big natural-uranium piles built during the war were too bulky to fit inside the hull of a submarine. Rickover sponsored a more compact design that used weapons-grade uranium rich in U235. The nuclear-powered U.S.S. *Nautilus* was launched in 1954. Nuclear submarines that carried ballistic missiles followed.

Technicians at the Tricastin nuclear power plant in southeastern France replace the uranium fuel elements in the reactor's core, an annual requirement for most nuclear power plants. Nuclear reactors create power by fissioning uranium atoms on a large but controlled scale. The resulting energy heats gas or liquid, which then heats water to form steam that drives the turbine on an electrical generator.

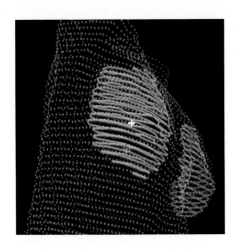

At the Los Alamos Health Research Laboratory, a plastic phantom (opposite) stands in for a job no human would perform: absorbing high doses of radiation. Precisely scaled and engineered, the model holds a solution chemically equivalent to human tissues. By bombarding the model with X rays, gamma rays, and neutrons, scientists can determine how the body absorbs and scatters radiation, whether from accidental exposure or from medical procedures, such as cancer therapy.

On a computer screen at Memorial Sloan-Kettering Cancer Center in New York (above), a "+" marks the precise spot on a breast tumor where radiation beams will be directed. This experimental computer technology uses three-dimensional images to bring new precision to the planning and delivery of radiation treatment.

Radiation therapy with X rays was born in 1896, just months after their discovery, when a Chicago medical student aimed the mysterious rays at a patient's breast tumor and the tumor appeared to shrink. Today, radiation is second only to surgery as a treatment for cancer.

Rickover also developed power reactors for aircraft carriers. In 1953 the U. S. Atomic Energy Commission authorized the first of those reactors converted to a civilian system for a power plant at Shippingport, Pennsylvania. The Shippingport plant was never economical without government subsidy, but it demonstrated a technology that would eventually account for more than 15 percent of U. S. energy output.

Another method of generating energy, controlled thermonuclear fusion, attempts to harness the reactions that drive the hydrogen bomb: the fusing of nuclei of light atoms. Thermonuclear fusion research has been going on since the early 1950s, with the prospect of commercial application in the 21st century. A practical fusion reactor would produce less radioactive waste than fission-powered plants and would run on a nearly inexhaustible fuel source—hydrogen from seawater.

One promising fusion technology uses powerful lasers to confine and heat a pellet of thermonuclear fuel. The laser has found many other less exotic applications since its first operation in 1960. As with nuclear energy, the laser demonstrates how scientists' pursuit of basic understanding of the physical universe can release a cascade of technological applications that alter and sometimes even revolutionize the world.

At the tiny scale of atoms, common sense is no longer a reliable guide. Many kinds of energy that behave like waves are actually particles. Microwaves and visible light are effects of speed-of-light particles called photons. Photons burst from atoms when the atoms are excited, as sound bursts from a plucked string, and their wavelength reveals where in the atom they originated. An atom produces a microwave photon or a visible-light photon when an electron orbiting its nucleus drops from a higher to a lower energy level.

A light bulb filament converts energy into light of many different and jumbled wavelengths—colors—that radiate out in every direction. A laser, by contrast, converts energy into pure, coherent light, all of one wavelength, that emerges in a narrowly arranged beam.

Albert Einstein deduced a basic principle of the laser in 1916. Exploring the fascinating interactions between light and matter, Einstein theorized that under certain circumstances light shining on matter could force it to emit additional light of exactly the same wavelength. This process of amplification, called stimulated emission, was confirmed experimentally in 1928.

The circumstances necessary for stimulated emission to occur were so unusual that scientists paid the phenomenon little attention until after the end of the Second World War. By then, the successful development of radar had increased interest in generating radio waves shorter than the microwaves that had made airborne radar systems possible. In all the wave generators devised by the late 1940s, the amplification of waves took place inside an electronic device—a vacuum tube—with critical dimensions similar in length to the waves being generated. There was a limit to how small such an amplifier could be made; centimeter-scale radar had already approached that limit.

Charles H. Townes, a tall, spare, South Carolina-born professor of physics at Columbia University, pondered the problem one morning in

Electrical discharges illuminate a tank at the world's most powerful particle accelerator, PBFA-II (left), built in 1985 at Sandia National Laboratories in New Mexico. Like its ancestor, the cyclotron of the 1930s, PBFA-II was developed to accelerate particles to high speeds, but not in order to fission atoms. The goal of PBFA-II is nuclear fusion—the joining together of atoms—to produce heat that could be used to power a conventional turbine generator. The self-perpetuating energy of stars, nuclear fusion has been heralded as the power source of the future: cheap, safe, and clean.

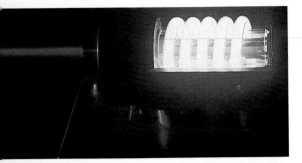

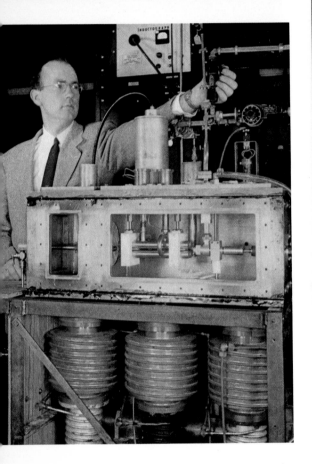

the spring of 1951 while sitting on a park bench in Washington, D. C. He had come to attend a meeting of a government committee which for two years had been discussing ways to generate waves in the short-millimeter wavelengths. "As I sat . . . musing and admiring the azaleas," Townes recalls, "an idea came to me for a practical way to obtain a very pure form of electromagnetic waves from molecules." Townes's idea involved stimulated emission—using atoms and molecules as amplifiers. They were certainly small enough. The problem was finding a way to assemble enough of them under the unusual conditions necessary for stimulated emission to work.

The way Townes found led to the development, in 1954, of the maser (the acronym stands for "microwave amplification by stimulated emission of radiation"). Masers eventually served as frequency sources of highly accurate atomic clocks and as amplifiers in radio telescopes.

A high-power industrial laser (left) cuts through steel to shape a saw blade. Albert Einstein discovered one of the basic principles of the laser in 1916, when he theorized that atoms could absorb light or other radiation and then be stimulated to release the borrowed energy. It wasn't until the 1950s that Charles H. Townes (far left), working with microwaves, hit upon a way to make a device based on this principle—a maser, precursor of the laser. In 1960 Theodore H. Maiman introduced the first true laser, made with ruby. Like most lasers, the ruby laser (left, upper) consists of three parts: a fluorescent substance—ruby, in this case; a source of energy—here a coiled flashtube; and a pair of mirrors, one on either side of the ruby rod. When excited by the energy of the flashtube, the fluorescent atoms in the ruby give off energy in units of light called photons. It is these photons, amplified and moving in synchrony, that produce the stream of pure red laser light.

Townes then began thinking about amplifying visible light. In the autumn of 1957, he teamed up with a Canadian colleague, Arthur L. Schawlow. The outcome of Townes and Schawlow's collaboration was a seminal scientific paper, "Infrared and Optical Masers," published in December 1958. There the two physicists laid out the basic theory of the optical maser—the laser, as it came to be called, the word "light" substituting for the word "microwave" in the original acronym—and described a possible system using potassium vapor as an amplifying medium. They thought that a very bright light could be shone on such a medium to "pump" its atoms to excited states that would prepare them for stimulated emission. They foresaw that setting the amplifying medium between two parallel mirrors would limit the direction in which the amplification could occur. If the mirrors were only thinly silvered, they suggested, the light as it reflected back and forth would eventually build to sufficient brightness to shine through the silver in a monochromatic, coherent beam.

Theory might be a general guide, but it did not lead Townes and Schawlow to build the first laser. A material they rejected as impractical served for the crucial invention. A young physicist at Hughes Research Laboratories in Malibu, California, Theodore H. Maiman, accomplished the historic breakthrough.

Maiman came late to the work. Many other scientists were trying to devise a laser, some of them with major government support, when he began his research in the autumn of 1959. "I like adventure," he reminisces. "This was the height of adventure. The competition was terrific." When most other researchers worked with gases, Maiman chose instead to work with a solid—a crystal of pure, man-made pink ruby. He had already constructed masers of ruby and knew the crystal's physical properties well. In gases, electrons can be excited to hundreds of different energy levels, which made experimenting with them complicated. Ruby, with only a few possible energy levels, was far simpler.

To make a laser, Maiman explains, "You need a lower level, a higher level, and a connection between the two. You need to be able to store more energy in the higher level than remains in the lower—an unusual condition called a population inversion. I liked ruby because it absorbs light in two broad bands and shifts to a fairly narrow intermediate band before it drops back to the ground state. There's an optical funnel, so to speak, from the broad blue and green bands over to the narrow red. That gives ruby an advantage. It takes a really intense light to pump a laser, and hot, bright lamps tend to radiate in a wide spectrum. If your material can only absorb a narrow part of that spectrum, it can't extract much energy from the light."

Electrons pumped up to the blue and green bands in ruby funneled quickly to the red band but then dribbled back spontaneously to their normal ground state. Maiman's bucket, in effect, had a hole in it. To achieve a population inversion, he had to pump electrons up faster than they dribbled back down. "My calculations showed I needed a light with brightness equivalent to about five or six thousand degrees Kelvin, which is about the brightest anyone's ever made a laboratory lamp. Then I remembered reading an article about the strobe lamps used in photography. It mentioned that strobes achieved temperatures up to 8,500 degrees Kelvin. They were pulse lamps. They just put out a burst. Everyone had been thinking of a continuous laser. But pulsed operation was just as good a place to start."

Maiman pored over the catalogs of flash-lamp manufacturers. "I found three lamps that had the temperature I needed. They were helical—coiled like a spring. I chose the smallest of the three. The ruby crystal was a little cylinder about three-eighths of an inch in diameter. It fit nicely right inside the lamp coil."

For parallel mirrors Maiman evaporated silver directly onto the flat ends of the little ruby cylinder. To allow the pulse of coherent light to escape, he scratched a small hole in one of the silver mirrors. Lamp and crystal fitted inside a reflective aluminum housing. Fully assembled,

At Walt Disney World in Florida, workers install fiber-optic cables that will transmit telephone and computer data by laser. Fiber optics—the delivery of laser light through strands of glass thinner than those opposite—has revolutionized the technology of moving words, pictures, and data. Countries from Great Britain to Indonesia now use optical communications systems. In most, lasers about the size of a grain of salt produce pulses of light from electrical current. These light waves, traveling through strong, flexible glass fibers, can transmit the entire Encyclopaedia Britannica *in a tenth of a second.*

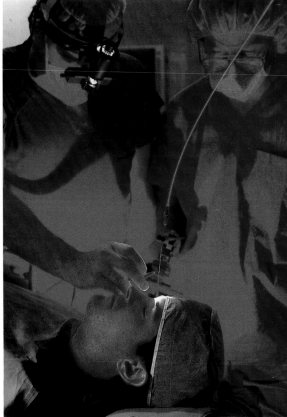

Powerful and precise, laser light has provided a versatile new tool for medicine. At Childrens Hospital in Los Angeles, a surgeon guides an argon-pumped dye laser through an optical fiber to destroy an eye tumor (above). Before surgery, the patient is injected with a light-sensitive dye that normal tissue excretes but cancerous tissue concentrates. When the red laser light strikes the cancerous tissue, the cells containing the dye absorb the laser's energy and are destroyed.

Though used primarily to treat eye problems, lasers assist in medical procedures from kidney-stone removal to throat surgery and cancer diagnosis.

the world's first laser was not much bigger than a flashlight battery.

"And then I set this up with my instrumentation," Maiman remembers, "and I fired it, and at a certain point it went." The physicist measured strong pulses of extremely pure red light from his elegantly simple arrangement on May 16, 1960.

Since 1960, materials of surprising variety have been made to lase. Other crystals followed Maiman's ruby. Gases came next. Physicist Ali Javan used helium and neon to create the first continuous laser in 1961. (All previous lasers emitted light in pulses.) Liquids were made to lase, as were dyes and even specially configured transistors. Incandescent light continues to serve as an energy source for lasers, but chemical reactions, beams of particles, even nuclear weapons have been used as well. Today masers and lasers operate at frequencies from microwaves through visible light all the way to X rays.

The powerful, coherent light that lasers generate has many uses. Laser surgery was an early application. Focused laser light cuts steel in heavy manufacturing. Laser gyroscopes keep commercial airliners on course. Laser radar makes precise measurements across great distances. In the laboratory the laser has become a tool as commonplace as the Bunsen burner used to be.

Perhaps the most ambitious use of the laser is in the research toward controlled thermonuclear fusion. The goal is to ignite thermonuclear fusion reactions that will release more energy than the lasers inject, ultimately to replace coal-fired plants and fission reactors as sources of power for electricity. In this work, two of the most surprising discoveries of 20th-century physics come together to demonstrate again what Albert Einstein's famous formula $E = mc^2$ first quantified in 1905: that matter is energy, vast almost beyond imagining, available for mankind's destruction or for its great and enduring benefit.

A scientist checks the alignment of a beam belonging to Nova (opposite), the world's most powerful laser, built in California in 1985 for experiments with nuclear fusion power. In theory, when Nova's beams strike a tiny fuel pellet made of hydrogen isotopes, the isotopes' nuclei fuse, creating a thermonuclear reaction. So far, however, fusion has proved impractical as a commercial energy source because power input exceeds output. Nova's beams are also directed toward making holograms of microscopic objects.

A hologram is created with a split laser beam, as a technician in France demonstrates (above left): The first beam illuminates a statuette before striking a photosensitive plate; the second beam strikes the plate directly. As the two beams meet at the plate, they create a pattern of light waves that, when illuminated by another laser, will produce a three-dimensional image. Invented in 1947, holography has a wide range of uses. At supermarket checkstands (above), laser scanners with holographic lenses read bar-code labels that identify products.

Computers & Chips

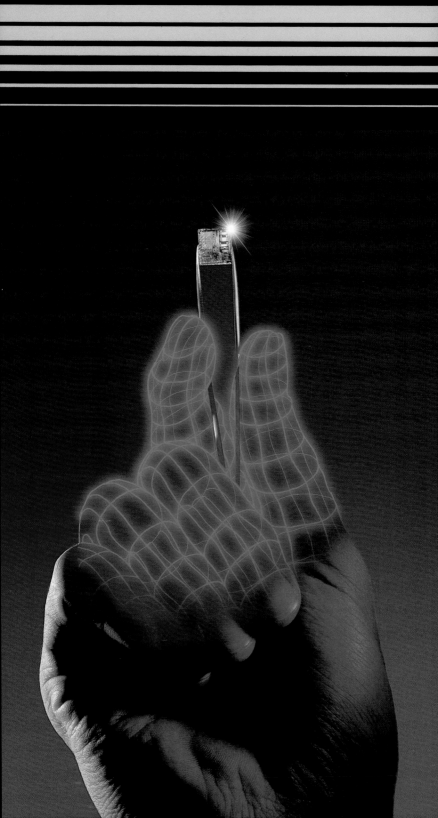

By T. R. Reid

Today we are living smack in the middle of a great technological revolution, just as profound as the industrial revolution. As steam and electricity expanded our physical powers, this new revolution frees us from mind-numbing computational labor and expands our intellectual powers.

Look up "computer" in any dictionary published before World War II, and you'll find something like this, from the 1933 edition of the *Oxford English Dictionary:*

> *Computer . . .* One who computes; a calculator, reckoner; *spec.*
> a person employed to make calculations in an observatory,
> in surveying, etc.

The definition refers to a human. Before the invention of the electronic digital computer and the microchip at its heart, some people spent lifetimes maintaining records and performing complex calculations that can now be completed with a few keystrokes.

As long as people have concerned themselves with engineering, census taking, navigation, taxation, astronomy, economics, weather forecasting, business, and other pursuits that require measurement and computation, they have been plagued by the need to carry out arithmetic calculations. Repetitive arithmetic is tedious, time consuming, and thoroughly unenviable work. For that reason, humans began early on looking for tools that could help. Ancient peoples used stones or sticks to help keep track of numbers. Our words "calculus" and "calculate" derive from the Latin *calculus,* a little stone.

One of the most venerable computing inventions began as a set of such stones or beads moved along lined boards and later mounted on a rack: the abacus. This ingenious, simple device, with separate rows of beads representing different orders of magnitude—ones, tens, hundreds, thousands, and so on—was probably developed in southwestern Asia more than 500 years before the birth of Christ. For more than two millennia it was the basic calculating tool of half the world. The Greeks and Romans used the abacus. Its popularity declined as that of Arabic numerals grew, but Europeans continued using it up to the 17th century or so. Traders probably took the abacus to China and Japan, and many of us think of it as an Eastern device because it is still in use there. With the advent of the pocket calculator, even the Orient is giving up the abacus—but not completely. The Japanese now make a pocket calculator built into the frame of an abacus. That way people who still don't trust the electronic device can check its answers.

Modern efforts to turn the drudgery of computation over to machines began with 17th-century European mathematicians. A German,

Wilhelm Schickard, devised perhaps the earliest mechanical calculator, but his so-called calculating clock was lost in a fire, and its design exists only in a letter he wrote to Johannes Kepler in 1624. The French prodigy Blaise Pascal was motivated to construct a "calculating box" in part by the overwhelming computational tasks required of his father, a local tax agent. In 1644 Pascal built a machine, similar in principle to Schickard's, that consisted of a series of connected rotary dials. Each dial had ten notches, marked 0 through 9. Turning a dial four notches would enter the digit 4; to add 3, the dial would be turned three more notches. A gearing system turned numbered cylinders to indicate a sum, 7. Gears also carried numbers from the ones column to the tens column, from tens to hundreds, and so forth. Pascal built a machine with eight dials, permitting addition and subtraction up to 99,999,999.

A generation later, the German Gottfried Wilhelm Leibniz developed a turning-dial machine that could divide and multiply as well. Leibniz, an inventor of calculus and one of the giants of mathematical history, wrote that he could not stand to watch fellow mathematicians "lose hours like slaves in the labor of calculation." So he improved the Pascal device, building a contraption in which gears turned cylinders bearing stepped ridges of graduated lengths. The Leibniz stepped wheel remained the basic mechanism for mechanical calculators until the mid-20th century, when mechanical devices gave way to electronics.

After the industrial revolution and the substitution of mechanical power for human brawn, another revolutionary idea came along: Perhaps machines could substitute for the human brain as well. This thought was the product of one of the most inventive minds of 19th-century England—that of Charles Babbage, a thinker so far ahead of his time that his greatest idea became a source of endless frustration.

Born in 1792, Babbage was a paradigm of the industrious English gentleman. He produced a long list of useful societal innovations, including the flat-rate postage stamp and the first reliable actuarial tables. A founder of the Royal Astronomical Society, he devoted countless hours to correcting the group's mathematical tables. Exasperated by one particularly long spell at this exercise in tedium, it is said, he burst forth with an idea that would become his life's work: "I wish to God," Babbage said aloud, "these calculations had been executed by steam."

Babbage tried to bring that wish to fulfillment with a Difference Engine—a gear-wheel mechanism that would calculate the tables using addition only. With large sums of his own and the British government's

Less than four feet tall and five feet across, the C-shaped Cray-2 supercomputer holds the power to unlock the mathematically complex secrets of everything from twirling tornadoes to colliding galaxies.

Since Blaise Pascal's adding machine of 1644, tools for computation have evolved from the primitive to the wizardly as mechanical gears gave way in turn to electromagnetic relays, *vacuum tubes, transistors, and integrated circuits. Computers grew big as houses before shrinking onto tiny silicon chips. The science of photonics may herald a new advance: Optical computers, which will use beams of light to achieve speeds perhaps a thousand times faster than those of their electronic counterparts.*

money, Babbage built a working model of the engine's calculating section. He soon became distracted with the idea of building a general-purpose calculating machine. The enormous Analytical Engine that he conceived—a calculator with numbers represented on thousands of geared wheels, powered by large steam engines—proved to be too intricate to build with limited 19th-century skills and tools. But Babbage's quest was not futile. Although ultimately frustrated, he pioneered fundamental ideas that have governed the design of computers ever since.

At the heart of the Analytical Engine, Babbage planned two wheel-filled sections: a "store," where numbers could be held while they awaited processing, and a "mill," where the mathematical work would take place. He further devised an apparatus to load numbers into the machine, and a primitive printer to show the results. In short, Babbage had established the essential four-part architecture still used in every computer: 1) an input device; 2) a memory unit (in England, this unit is still called the store); 3) a central processor, Babbage's mill; and 4) an output mechanism.

Another Babbage innovation was his ingenious plan for loading data and operating instructions into his computing engine. For this task, Babbage borrowed an idea from the French weaver Joseph-Marie Jacquard. Jacquard had devised a way to automate weaving with a series of punched cards that would dictate the pattern on the loom. Babbage, who was interested in weaving as well as in everything else, learned of Jacquard's method and adopted it to encode instructions for his Analytical Engine. As Babbage's chief assistant and publicist explained the idea, "The Analytical Engine *weaves algebraical patterns* just as the Jacquard loom weaves flowers and leaves."

That assistant was Augusta Ada Byron, daughter of the poet Lord Byron and wife of Lord Lovelace, a proper British gentleman who was, fortunately, not quite so proper as to forbid his wife from associating with Babbage and his strange projects. A natural mathematician in a world where math was considered the province of males, Lady Lovelace understood Babbage's calculating engine and took on the task of writing explanations about it and how to use it—explanations so good that many have credited her with inventing the programming principles herself. To honor her work, the Pentagon has given her name to its standard programming language, ADA.

Some of the concepts advanced by Babbage and Lady Lovelace were put to use a few decades later in a mechanism called the Hollerith Machine, after its inventor, the American engineer Herman Hollerith. As usual, the impetus for the new computing device was an overwhelming computational task—in this case, the upcoming 1890 United States census. Always an enormous accounting chore, the job of counting and analyzing the 1880 census had taken a large corps of human computers more than seven years. Unless something were done to automate the process, the 1890 census calculations and analysis might still be unfinished when the 1900 census was due to begin.

Hollerith solved that problem with an electrically powered high-speed tabulator that used the Jacquard/Babbage input mechanism—

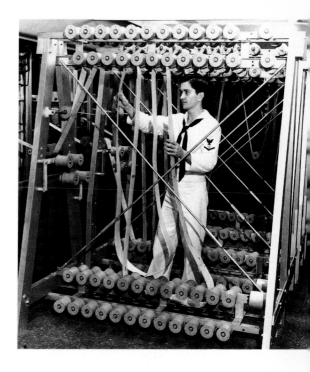

Searching for ways to feed instructions to his Analytical Engine, 19th-century Englishman Charles Babbage took inspiration from the Jacquard loom (opposite), which used punched cards to guide the weaving of intricate brocades and damasks. By pressing spring-loaded pins against a series of cards, the loom activated hooks that lifted warp threads corresponding to holes in the cards. Some computers still use a similar approach, reading punched cards with wire sensors that complete electrical connections by contacting metal plates or rollers behind the holes.

Punched paper tape (above) programmed the 51-foot-long Harvard Mark I, an electromechanical device developed by Howard Aiken and a team of IBM engineers to compute ballistic tables for the United States Navy in the 1940s.

John von Neumann

One day early in this century, the legend goes, young Jancsi von Neumann sat with his mother in the parlor of their Budapest home. When she paused in her crocheting and stared into space for a minute, her son inquired, "What are you calculating?"

Even as a child, John von Neumann—whose name would become synonymous with computer design—immersed himself in a life of the mind that held no pursuit more enjoyable than calculation. At six he could divide two eight-digit numbers in his head; at eight he conquered calculus; by eighteen he had published original work in mathematical theory. But this life did not begin and end with math. In a colleague's words in 1958, "To follow chronologically von Neumann's interests and accomplishments is to review a large part of the whole scientific development of the last three decades."

Von Neumann loved history too, poring page by page through the 21 volumes of the Cambridge ancient and medieval series. A photographic memory kept knowledge handy he had acquired years before. Asked as a party joke to recite *A Tale of Two Cities*, von Neumann plunged in: "It was the best of times, it was the worst of times, . . . " Fifteen minutes later his astonished taskmaster called a halt to the recital.

The Budapest into which Jancsi was born in 1903 was a city of precocious genius. Contemporaries included Edward Teller, Eugene Wigner, and Leo Szilard, all, like von Neumann, future brainpower for nuclear arms development in the United States.

Yet Jancsi's special gifts stood out. He simultaneously pursued two degrees, a diploma in chemical engineering from Zurich and a doctorate in mathematics from Budapest, and spent spare time in Berlin on more

Mathematical and computer genius John von Neumann gives a lecture in the 1950s.

study. In 1927 he began teaching in Germany and worked on the amalgamation of math and physics known as quantum mechanics—the backbone of nuclear physics. There he took the honorific "von" as an equivalent to a minor Hungarian title held by his father Max, a prominent Jewish banker.

Von Neumann's teaching style reflected the computerlike nature of his brain. He would race along in a proof, light-years ahead of his students, erasing notations before the hapless pupils could absorb them.

A visiting appointment at Princeton in 1930 later turned into an invitation to join the new Institute for Advanced Study. At 29, von Neumann found himself along with Albert Einstein on the think tank's original roster.

In Princeton, dapper and engaging "Johnny" von Neumann came into his element. He and his current wife (he married twice) threw legendary parties where guests were treated to fine food, drink, talk—and samples of their host's vast repertoire of bawdy limericks. But too much play made the host anxious, and he would abruptly retreat to his study to think out a problem, one which, as war approached, often related to his adopted country's defense.

Von Neumann joined the war effort as a ballistics consultant, spending

A test launch of a 1980s Trident 1, successor to ballistic missiles that von Neumann's computers helped control.

much time at the Aberdeen Proving Ground in Maryland. His call to nuclear arms came in 1943 when J. Robert Oppenheimer invited him to Los Alamos. A mathematician who could think and talk like a physicist was ideal for the Manhattan Project. With characteristic logic and clarity, von Neumann mapped out the calculations that governed a nuclear explosion.

During a trip to Aberdeen he learned of the ENIAC computer under construction at the Moore School in Philadelphia and visited there. Quickly absorbing the essence of electronic computers, von Neumann synthesized the guiding logic of computer architecture. Though not all of his ideas were original (many were presaged by 19th-century visionary Charles Babbage), they were exceptionally well presented. Today, computers that incorporate the elements von Neumann propounded—central arithmetic unit, central processing unit, memory storing both data and instructions, input and output units, use of binary numbers, Boolean logic, and serial processing—are still called von Neumann machines.

The horror of Hiroshima and Nagasaki did not deter von Neumann, as it did some of his colleagues, from advancing the cause of nuclear energy. He stayed on at Los Alamos to work on the hydrogen bomb and eventually landed a top position on the Atomic Energy Commission. Seeing the intercontinental ballistic missile (ICBM) as America's hope in the arms race, he pressed for the computer power to ensure successful deployment.

Von Neumann's extraordinary mind may have fallen short in its calculation of the risk to himself from radiation. He witnessed a test in New Mexico and, like others who helped unleash the atom's power, may have become its victim. He died of cancer in 1957, surrounded by attendants cleared for any unintentional utterances of top secret information stored in the computer that was John von Neumann's brain.

Catherine Herbert Howell

punched cards. Information from the cards was tabulated by a series of counters with clocklike dials, from which totals could be copied by hand. The new, mechanized census took less than three years to analyze data about the 62,622,250 people living in the U. S. in 1890.

Hollerith's machine was basically just a sorting and tabulating tool. By the mid-20th century, progress in fields like chemistry, physics, astronomy, meteorology, communications, and aviation posed formidable new requirements for complex calculation. For example, faster and more powerful weaponry forced the U. S. War Department to form "computer pools," where scores of people worked month in and month out computing trajectory tables to be used in aiming big guns. The burgeoning thirst for computational power prompted an era of breakthroughs that led to the computer revolution.

The first stage of this era, roughly from 1930 to 1950, peaked during the war years. Some of the work occurred overseas. In Berlin the young German engineer Konrad Zuse built and programmed three working computers—only to see his plans squelched by the Nazis, who thought they would win the war without him.

But they didn't, partly because of British efforts in the same field. Working in secret at Bletchley Park near London, Britain's top mathematicians and engineers hurriedly built an electronic computer series called Colossus to replace the electromechanical machines that were already succeeding in cracking Axis codes.

In the U. S., meanwhile, much of the computer pioneering work centered on a "fertile crescent" running from the laboratories of the Massachusetts Institute of Technology and Harvard in Massachusetts, to Bell Telephone Laboratories and the Institute for Advanced Study at Princeton in New Jersey, then to the Moore School of Electrical Engineering at the University of Pennsylvania. Two new machines constructed in the crescent were the Differential Analyzer at MIT, which Vannevar Bush developed in 1930, and the Harvard Mark I computer, designed for IBM in the 1940s by Howard Aiken.

Both were latter-day Babbage engines, complex machines with shafts and gears powered by electric motors, but they differed from each other in a fundamental way. The Mark I computed with discrete numbers, a method of calculating derived from our basic counting tool: the human hand, with its eminently countable fingers. Any machine that works by counting individual items—be they abacus beads or electronic pulses—is a "digital" device (from the Latin *digitus*, finger).

The Whirlwind, an electronic, digital, room-size computer, was built by Jay Forrester and others at MIT in the mid-1940s. Designed to solve problems of aircraft control in simulated flights, it quickly found other uses in scientific research and helped modernize U. S. air defenses and air traffic control. The Whirlwind employed probably the earliest versions of modern video display terminals.

In the late 1950s Robert Noyce of Fairchild Semiconductor (above, foreground, with colleagues) and Jack S. Kilby of Texas Instruments independently developed the first integrated circuits. They built transistors and other components onto a single piece of silicon—a "chip." Soon the race was on to cram smaller components onto each chip and thus boost speed and capability.

Texas Instruments in 1983 posed a ladybug on a silicon wafer (opposite, upper left) and magnified them both 6, 12, 45, 448, and 3,600 times to show that the electronic conductors on its computer chips are as thin as a hair on an insect's foot—about one ten-thousandth of an inch wide.

Bush's invention, however, represented numbers with a range of continuous positions. In the Differential Analyzer, a disk rotated halfway around might be an analogue for the number 5—just as the minute hand moving halfway around the watch dial is an analogue for the passage of half an hour. In theory, a disk with an infinite number of positions can count in infinitely small increments. Clocks and calculators that use physical measurement—of rotation, of length, of voltage—as analogues for numbers are "analog" devices.

Both of these new mechanical computers worked, but were so big and so prone to breakdowns that it became clear a different course was required. A new design idea emerged—to represent numbers not with cumbersome mechanical gears or relays but with arrays of electronic switches, such as vacuum tubes.

In 1945 a fully electronic, general-purpose digital computer was completed at the Moore School in Philadelphia. Its inventors were Penn instructors J. Presper Eckert and John W. Mauchly. (The two later lost a heated patent dispute in which the court gave credit for some key concepts to John V. Atanasoff of Iowa State University, whose earlier prototype machine Mauchly had seen and discussed.) The Electronic Numerical Integrator And Computer was better known as ENIAC. ENIAC was a leviathan; it occupied a thousand square feet of floor space; it had 18,000 vacuum tubes and its own air conditioner, which struggled to keep the machine cool enough to run. Eckert and Mauchly designed it for U. S. Army Ordnance to calculate firing patterns. But ENIAC proved capable of many other types of calculation. It was the fastest, most versatile computer yet, and it paved the way to the future.

Meanwhile, computer pioneers were developing a set of principles that would govern modern computers. John von Neumann, a revered

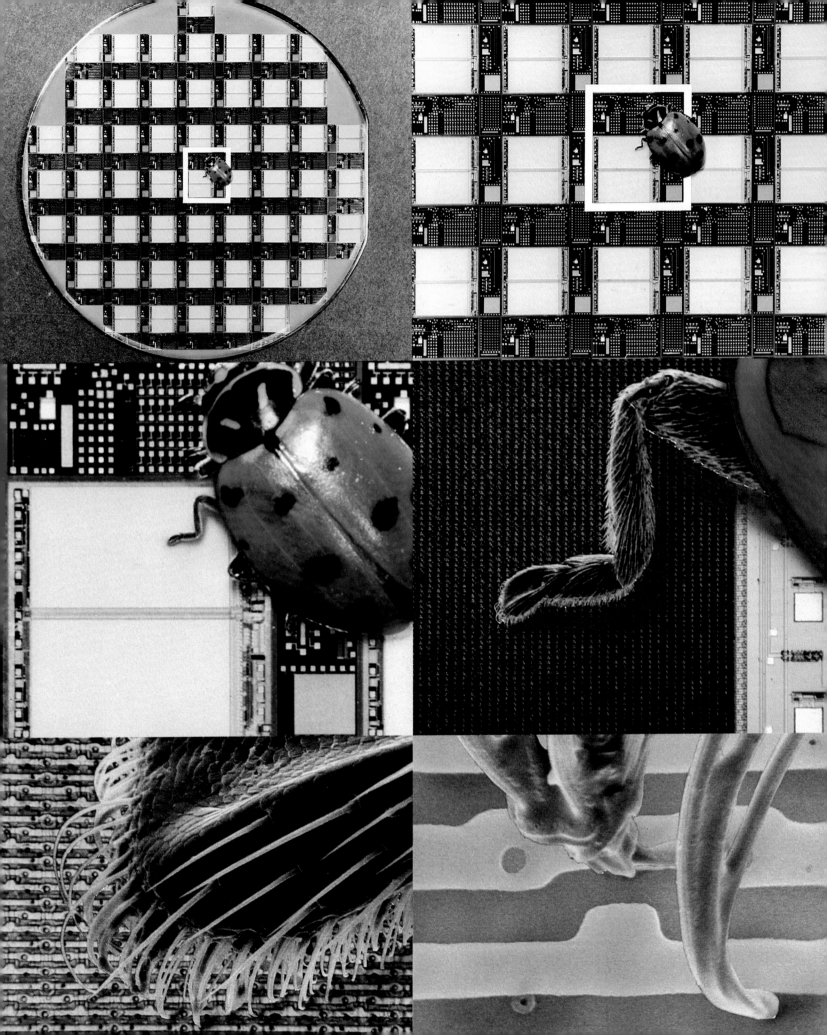

258

mathematician at Princeton's Institute for Advanced Study, and Alan Turing, a brilliant British logician who worked on Colossus, were among those who argued that computers should not use the base-ten, or decimal, number system that we learned in school. A computer could function faster and more efficiently if it used the simplest system of all—the base-two, or binary system, in which all numbers are represented by combinations of just two digits, 0 and 1. This way you can represent 0, 1, 2, 3, 4, 5, 6, for instance, as 0, 1, 10, 11, 100, 101, 110. When the computer finished working a problem, it could convert the binary answer back to decimal notation for human consumption.

Binary math meshed nicely with the use of electronic switches. A switch can take only two different positions: On or Off, perfect for representing binary numbers, which need only two characters, 0 and 1. Thus chains of electronic switches inside a computer could represent any number digitally. They could flash through calculations fast as blinking lights.

To complement binary numbers and binary switches, an MIT graduate student, Claude Shannon, had worked out a system of binary logic to permit computers to make comparisons and decisions and thus run through a "program," or sequence of operations, without constant human intervention. Shannon based his computer logic on a monograph written in the 1840s by a visionary English mathematician, George Boole. Boole had devised a complete system of logic in which any decision, no matter how complex, could be reduced to two variables, Yes or

As circuits on computer chips evolved from the simple mazes of the 1960s (opposite, upper) to the labyrinths found today (opposite, lower), the manufacturing process also became more sophisticated. In rooms swept clean by the flow of filtered air, workers grow silicon crystals and slice them into wafers. After insulating, coating, and masking the wafers and exposing them to ultraviolet light, the workers place them in ovens to be etched by superhot gases (below left). They then dope them with impurities, dice them into chips, and house them in lead-wire frames (below). A new material, gallium arsenide, promises chips ten times faster than silicon.

No. Boole's construct received only limited attention in his lifetime; its practical application for computation had gone unrecognized for nearly a century when Shannon pointed out that Boolean logic was perfect for binary computers.

The first full-scale machine designed to incorporate these principles—binary numbers, vacuum-tube switches, and Boolean logic—was EDVAC (Electronic Discrete Variable Computer) at Philadelphia. EDVAC was also the first U. S. computer designed to store its own programs. This breakthrough, which the British had made two years earlier, meant that now a computer could hold instructions for its various operations in its memory and call them up in a fraction of a second. The program could interact with other programs and even modify them—transforming the computer into an almost infinitely flexible machine.

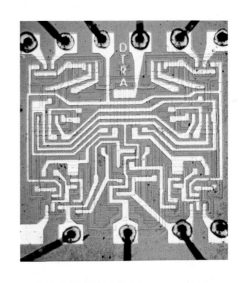

A host of similar machines followed, each offering improvements in speed or circuitry and each boasting its own fancy acronym. In the decade following the end of World War II there were the EDSAC (Electronic Delay Storage Automatic Calculator) at Cambridge University in England, the MANIAC (Mathematical and Numerical Integrator And Computer), and even the JOHNNIAC, named to honor von Neumann.

The best known of these early vacuum-tube digital computers was the UNIVAC (Universal Automatic Computer), built by Eckert and Mauchly in 1950. UNIVAC soared to overnight fame on election night in 1952. On a CBS News broadcast with Walter Cronkite, it predicted the election of Dwight D. Eisenhower based on a mere seven percent

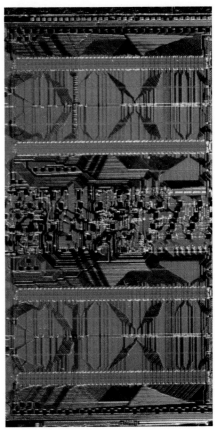

of the vote—hours before the official tabulations were available.

Actually, UNIVAC didn't predict anything. The only thing it did—the only thing it could do—was calculate a series of complex equations that a team of engineers and political scientists had devised beforehand. The election-night triumph really belonged to the programmers who told the computer what to do.

The only thing computers do is follow instructions—and they can be maddeningly literal about it. A new programmer usually learns that lesson the hard way. He or she writes a simple program to, for example, "find the square root of 562,000." The computer appears to do nothing. The befuddled programmer eventually figures out that the computer did calculate the square root, just as it was told. But it did not show the answer on the screen; that step wasn't in the program. The programmer just assumed that the machine would be smart enough to report the answer. But computers assume nothing. They do precisely what they are told.

Lady Lovelace realized this clearly. "The Analytical Engine," she wrote, "has no pretensions whatever to originate anything. It can do whatever we know how to order it to perform." That basic principle made things difficult for the small band of engineers and mathematicians who developed the science (or, as some say, the art) of computer programming during and after World War II. Those early programmers had to master a new syntax of communication, a syntax that assumed nothing and spelled out each step required to solve every calculation.

At the Winnebago factory in Forest City, Iowa (opposite), a network of terminals hooked to two IBM 4381 mainframe computers enables workers to design new motor home models and control inventory. Nearby, a farmer can check the latest pork-belly prices from a printout received via satellite dish, while students study in the school's computer learning center. In the early 1980s Control Data Corporation chose Forest City as a prototype for a system that would link the schools to a library of educational software in Minneapolis. With a grant from Winnebago's founder, Control Data brought high technology to this town of 4,500 people.

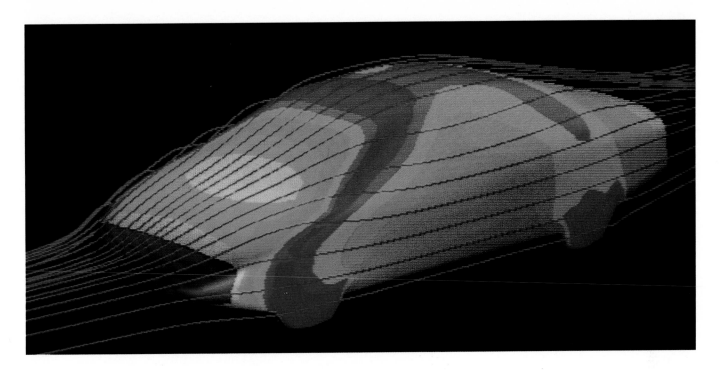

Furthermore, they had to converse in language that machines could digest. Once the binary system was adopted, programmers spent their days writing instructions like "11101011 1010010 1001110".

One programming pioneer was Grace Murray Hopper, a math professor at Vassar and Barnard Colleges. She joined the Navy Reserve in 1943 and was assigned to help write instructions for the Mark I at Harvard. For the next 30 years, in and out of the Navy, she played a key role in programming many computers, including the UNIVAC that hit the bull's-eye on election night. Among programmers, Hopper is most respected for compiling the first "high-level language"—that is, a programming language that permits people to talk to a computer in terms close to human dialect: "IF X > 0, THEN PRINT X/2."

When Hopper began work on her first high-level language in 1951, the project was generally considered superfluous. What need was there for a computing language that ordinary people could understand? After all, vacuum-tube digital computers were huge, expensive, power hungry, and few. The world needed only a tiny tribe of programmers to instruct them. It was commonly thought in the 1940s that computing would be restricted to numerical calculation and that four or five

Today's supercomputers help design the vehicles of tomorrow. At Cray Research in Chippewa Falls, Wisconsin (opposite), a technician tightens a panel on a sleek Cray-2 supercomputer, which can perform more than a billion operations per second. The super speed comes from closely packed chips immersed in fluid to keep them from overheating. NASA scientists rely on a Cray-2 to test aircraft and spacecraft designs on the drawing board. Supercomputer data helps create images that show how various designs would affect airflow. General Motors also uses data from a Cray to study the aerodynamics of auto designs (above). The technology is moving so fast, one expert estimates, that in five years a supercomputer should be able to do in one day what now takes it three months.

Caught in an electronic web, traders unleashed panic on the floor of the New York Stock Exchange in October 1987, as computerized trading helped push stocks to their greatest single-day loss. Computers monitoring giant portfolios were programmed to sell as prices slid to certain levels. The volume of that sell-off sent prices still lower. Fearing the worst, investment fund managers leapt to their consoles to dump billions of dollars worth of stocks, triggering more rounds of programmed selling. Thus the slide fed on itself, pushing the Dow Jones average down more than 20 percent. Shock waves coursed through the electronic network linking the world's financial centers. Automatic trading halts on the Tokyo Stock Exchange (above) helped cushion that market's fall.

high-speed computers, strategically located, would fill all the world's calculating needs. This notion was implicit in George Orwell's novel *1984*, published in 1949, which envisioned a world where a few giant bureaucracies employed massive computers to enslave the population.

But in reality the commercial and industrial uses of computers were already apparent. Companies such as Remington Rand—which had bought out Eckert and Mauchly and continued to build the UNIVAC— and IBM jumped headlong into manufacturing business machines. They continued to improve circuit design, computer memory, and the ways data could be fed into and retrieved from the machines.

The Orwellian vision receded further when two new inventions led to almost unthinkable reductions in the size and cost of computers. Although a great improvement on mechanical machines, computers that calculated by switching vacuum tubes on and off had their own disadvantages. Some work had been done to miniaturize them, but the glass tubes were still big, expensive, unreliable, and power hungry. When thousands of tubes were operating close together, they generated so much heat that cooling them down was a serious problem. (One computer was used in winter to heat the entire building it occupied.) All these problems set a limit on size. There was no point in designing a computer with 100,000 vacuum tubes if it was likely to reach meltdown temperature after it was turned on.

The first breakthrough occurred two days before Christmas in 1947, when a team of physicists at Bell Telephone Laboratories demonstrated a new kind of electronic switch: the transistor. Made of a solid piece (hence the term "solid-state") of a semiconductor such as silicon, this device could switch current on and off even faster than a vacuum tube—and do it with little heat, without a vacuum and the glass bulb to maintain it, and with minute amounts of power. The transistor brought the cost of computers down and led to the development of minicomputers and of powerful new mainframe computers such as the IBM 360.

A new world dawned for computer designers. Using transistors, engineers could now draw up blueprints for computers with 100,000, 500,000, even a million switches, all taking up less space and using less

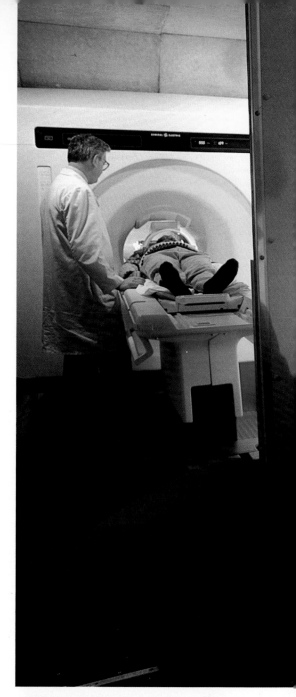

A computer in Minnesota's NetCare mobile testing unit teams up with a powerful magnet and a radio transmitter to reveal the inner workings of the body (right, upper). Computers analyze data from a magnetic resonance imaging machine (MRI) to create detailed pictures of body structures and chemical compositions. The process makes the watery tissues in the brain show up clearly and can uncover an array of problems, including brain damage and nerve degeneration.

Hoping to solve the mystery of human consciousness, scientists at EEG Systems Laboratory in San Francisco have used computers to learn how different parts of the brain interact. First, an MRI scan depicts cross sections of a person's head with electrodes marked for reference (right). Next, the computer assembles several scans to make a three-dimensional view (opposite, left). Finally, neural-network software incorporates the results of an electroencephalogram (EEG) in a computer model (opposite, right), using colored lines to reveal brain activity as the person prepares to play a video game.

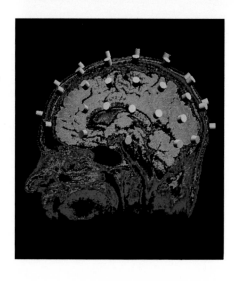

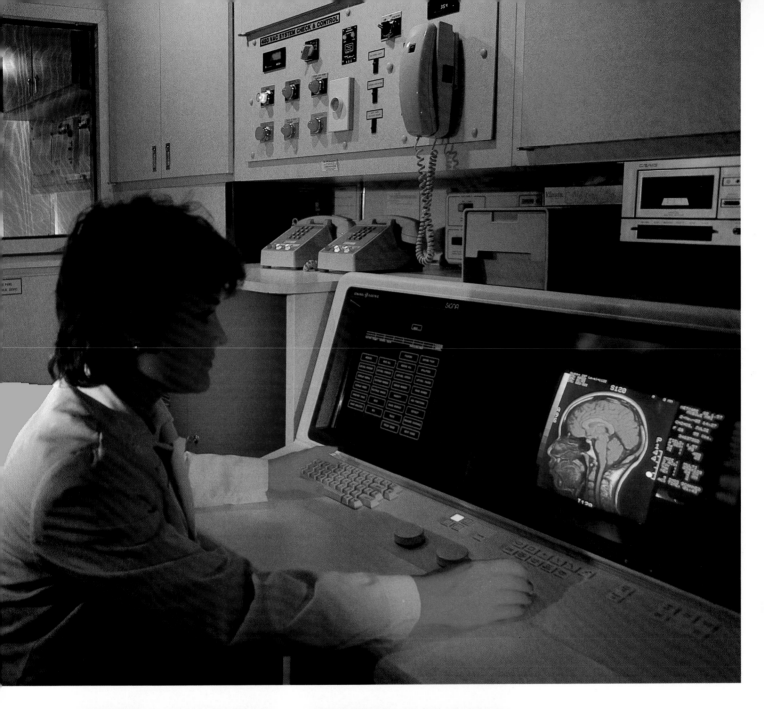

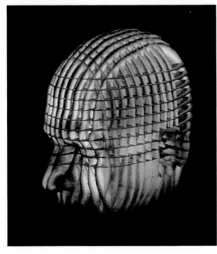

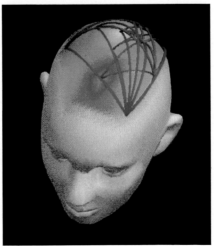

power than a vacuum-tube counterpart with a tenth of the capacity.

But now another obstacle emerged. It was one thing to design circuits with millions of components—it was something else to build them. Humans equipped with lengths of wire and soldering tools would have to construct millions of intricate connections by hand. There was almost no way that humans could construct several million of anything without making a few errors along the way. And even one faulty connection could render an entire computer inoperative.

The "tyranny of numbers," as computer engineers called this problem, was a near-total block to technological advance. By the mid-1950s the U. S. military and the nascent American space program needed computers powerful enough to steer missiles and moon rockets through the stratosphere. They funded major research efforts. All over the world, scientists and engineers searched for a solution.

The daring idea that conquered the tyranny of numbers was developed independently by a pair of Americans: Jack S. Kilby, of Texas Instruments, and Robert Noyce, of Fairchild Semiconductor.

Kilby, a quiet, introverted, deliberate thinker who likes to ponder a problem alone, hit on the concept first, in the summer of 1958. Instead

A computer-driven robotic hand flexes its fingers with the aid of tiny pneumatic devices. Built by researchers at the University of Utah and MIT, the hand is dexterous enough to gently crack eggs. Such devices someday may work on precision jobs in industry.

Inventor George Devol, Jr., patented the first programmable, automatically controlled robot in the 1950s and, with engineer Joseph Engelberger,

began manufacturing industrial robots. Sales did not take off until the late 1960s. Today some 300 firms worldwide build robots, and more than 100,000 "steel-collar" workers perform onerous jobs such as toting or spray painting. In Canada robots weld Dodge and Plymouth mini-vans on a Chrysler assembly line (above).

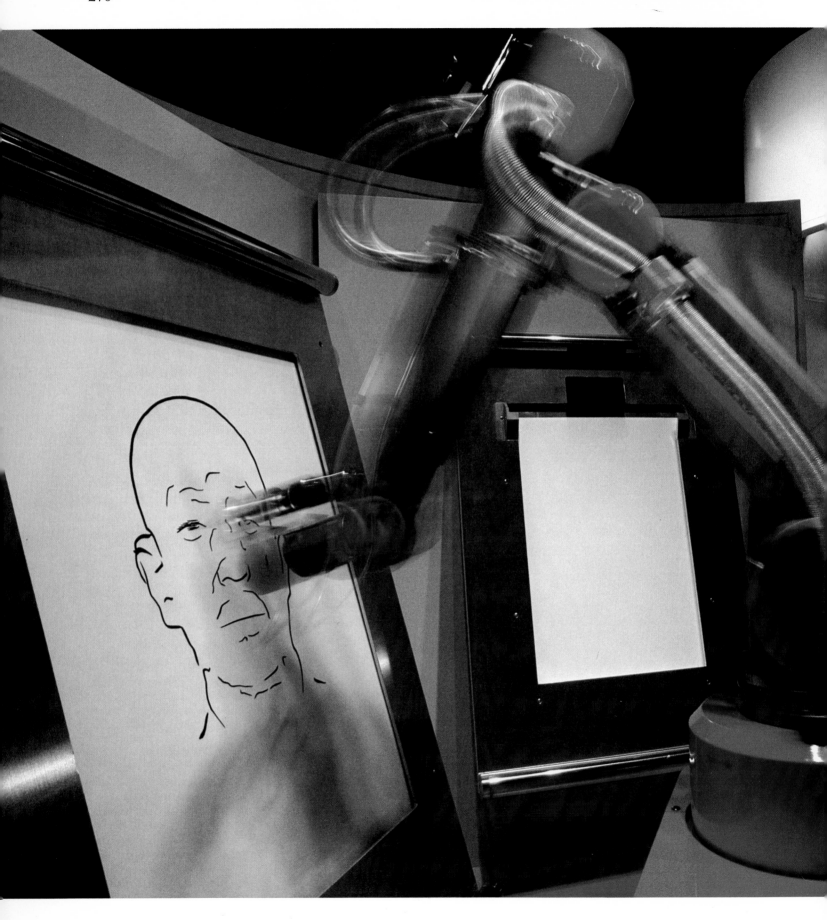

of building circuits from millions of individual components, Kilby suggested that the components of a circuit—transistors, as well as resistors, which block current, and capacitors, which store a charge—could be manufactured together as part of a single chip of germanium or silicon. He built a crude prototype on a germanium wafer, jury-rigging the circuit's connections with tiny wires.

Robert Noyce, an outgoing, talkative, impulsive inventor who says he needed to work with colleagues so they could sort out his good ideas from his bad ones, arrived at the same conclusion about six months later. Noyce was then an officer of Fairchild, one of the first high-tech firms to locate south of San Francisco in the sunny farm country that is known today as Silicon Valley. In the business of building transistors, Fairchild had developed a process that would permit a number of them to be made on the same side of a piece of silicon. While batting ideas back and forth with a colleague, Gordon Moore, Noyce suggested that the same process could be used to link numerous transistors and other components into a circuit without using wire. And such a circuit, integrated into a single chip of silicon, would solve the tyranny of numbers.

Noyce and Kilby, whose impact has led them to be called "the Henry Ford and Thomas Edison of our times," went ahead independently to develop the tiny integrated circuits, or "chips," they had conceived. Their invention changed the face of computing. Enormously cheaper and smaller machines meant the computer could move beyond multi-million-dollar laboratories and become a tool for everyone. Over time, engineers learned how to cram more and more transistors—and thus more and more computing power—onto the microchip. Today's advanced chip designs provide as many as 35 million transistors on a flake of silicon thinner than a sheet of paper.

In 1971 a colleague of Noyce's, Marcian "Ted" E. Hoff Jr., designed a chip that contained all four essential parts of a computer: input circuitry, memory, processing unit, and output circuits. This "computer-on-a-chip," a microprocessor, made it possible to computerize ordinary tools and appliances. Today there are chips by the dozen in countless American homes and cars—in digital clocks and calculators, toasters and typewriters, thermostats and TV sets. ENIAC, the 1940s behemoth, filled a room and consumed enough electric power to drive a locomotive. A modern microprocessor, smaller than a thimble, can run on penlight batteries and outcompute at least a thousand ENIACs.

Soon after Hoff developed the microprocessor, electronics hobbyists

An artistic robot puts pen to paper at a fair in Japan. After a camera scans the subject's face, the robotic hand whirs into action to sketch a likeness. Driven by computer, the robot converts light intensities into numbers and analyzes them to guide the pen. Robotic vision may take over such detail jobs as inspecting computer chips and make possible mobile, autonomous robots for home and industry.

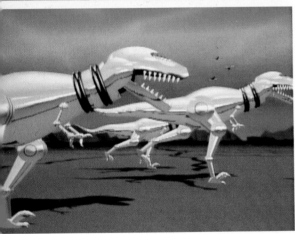

began wiring keyboards, television screens, and microprocessors together into small, simple home-brew computers. Among the dedicated techies pursuing this hobby were two young Californians, Stephen Wozniak and Steven Jobs, who built small computers in a garage. In 1975 the skillful Wozniak designed a machine that was cheaper to build but still more powerful than most other homemade computers. Wozniak hoped to sell a few models to fellow hobbyists. But Jobs had much more ambitious ideas. He thought the personal computer could have universal appeal if it did not appear so technically forbidding. So Jobs gave his computer the least forbidding name he could think of—Apple—and set out to get it into every office, shop, and rec room in America. By its tenth birthday, Apple Computer was a billion-dollar-plus success story, and personal computers were selling at a rate of ten million per year in the United States alone.

Today computers range in size and purpose. Minute microprocessors embedded in the human body can regulate damaged organs. Powerful, high-speed supercomputers, such as those designed by American engineer Seymour Cray, can process the enormous amounts of data involved in defense work, weather forecasting, satellite mapping, and a rapidly growing number of other fields. Computers are in use in the home and in the hospital, in the store and in the school. Computers read books to the blind (in the listener's choice of 50 simulated voices) and convert the spoken word into print for the deaf. Computers manipulate information in every form: not just numbers but also words, pictures, charts, photographs, maps, and sounds. Computers may even ease visits to the dentist: With systems now in development, a dentist will be able to have a computer design and manufacture a crown simply by scanning the needy tooth with a laser.

More than laborsaving calculators, computers have opened new realms of science, engineering, and art. They help us see the faces of distant planets, publish magazines from a desktop, design buildings in three dimensions, paint pictures with electrons, and probe the nature of the universe. And, of course, they help us invent new computers. The computer is now of such broad utility that the literal meaning of its name is dwindling into language history:

Computer . . . one that computes; *specif:* a programmable electronic device that can store, retrieve, and process data.
—*Webster's Ninth New Collegiate Dictionary*, published 1984.

To pull a robot out of a hat for a magazine cover (opposite), artist Richard Chuang used computer magic with software developed by Pacific Data Images of California. After sketching the picture by hand, he fed mathematically coded information about the design to a Ridge-32 computer, which created a three-dimensional model on its display screen. He then refined and colored the picture and recorded it on film. For the film Chromosaurus *(left), animator Don Venhaus used similar techniques to form dinosaurs with gleaming hides. He laboriously plotted the motion of each dinosaur on a computer and then filmed the animals in action. In the finished product they move 60 times a second.*

Engineering Life

By H. Garrett DeYoung

On May 14, 1796, the first act of a fateful drama unfolded in a village in Gloucestershire, England. The cast consisted of Edward Jenner, a doctor, natural historian, and poet; an eight-year-old village boy named James Phipps; and Sarah Nelmes, a young milkmaid infected with cowpox, a disease of dairy cows that is also contagious to humans. Jenner gently pricked some fluid from a sore on the girl's wrist with a lancet, then scratched the fluid into the boy's arm. As expected, James soon came down with a mild headache, chills, and other short-lived symptoms of cowpox.

Six weeks later Jenner again inoculated James Phipps. This time he used a drop of what he called "variolous matter" taken from a victim of smallpox, the dreaded scourge that killed one out of three 18th-century English children. James Phipps never developed smallpox, even after another exposure. Jenner later wrote that the boy remained healthy with "no sensible effect . . . on the constitution."

Inoculation was not a new idea in 1796. It had been used for decades in Europe, and much longer in Asia, to ward off smallpox. But the traditional method—dosing with a mild form of the disease—generated fear and censure, because mild smallpox could turn virulent.

Although his method worked, Jenner could only speculate about the similarity between the relatively harmless cowpox virus (the so-called variolous matter) and the human smallpox virus. Neither did he understand the biological mechanism by which inoculation with one produced a lifelong supply of antibodies against the other. (Indeed, the idea of antibodies—proteins created by the immune system to destroy foreign organisms—would not be proposed for another hundred years.) But Jenner's meticulously recorded experiments showed that resistance to a disease could be induced with a vaccine (from the Latin *vacca* for "cow"), a mild variety of the infectious matter.

With his vaccine, Jenner proved scientifically that a substance could be used to help the human body by influencing its functions. Inventors had long ago learned to manipulate the inorganic world with machines, most recently those powered by steam. Scientists from Jenner on would increasingly find human life, too, amenable to change. From their discoveries would spring a plethora of drugs—both natural and synthetic—that would not only prevent but also diagnose and treat disease. These drugs would profoundly change the course of medicine.

Jenner's results were not adopted by the medical profession for 90 years. The adoption was spearheaded not by a physician but by a French chemist named Louis Pasteur, who was the first to realize that vaccination might also be used against other diseases.

In the 1840s a professor fired Pasteur with enthusiasm for chemistry, and soon he was established as one of France's most innovative

chemists. His studies for the French wine and beer industries on the link between microorganisms and fermentation led him to espouse a hotly disputed idea. The "germ theory" held that diseases in humans and animals did not arise spontaneously from the body, as had widely been supposed, but were caused instead by microorganisms, which could transmit the disease through air and food. If that were the case, reasoned Pasteur, infection could be prevented by destroying these microbes, these "little nothings that . . . carry disease and death."

Even though infection was rampant in hospitals then, most physicians doubted that this smug chemist was likely to provide the remedy. But in Scotland a 33-year-old surgeon named Joseph Lister read about Pasteur's theories and decided to test them. He began using carbolic acid to disinfect patients' wounds, his hands, and his instruments. By the late 1860s the death rate from infection in Lister's Glasgow Royal Infirmary operating room had dropped from 45 to 15 percent.

Pasteur went on to identify the bacterium that causes the deadly livestock disease anthrax and, in so doing, removed any doubt that germs cause disease. He then created an anthrax vaccine from anthrax bacteria that had been bred to weaken their virulence. In a public demonstration, he vaccinated part of a herd of sheep, then injected the entire herd with anthrax. As Pasteur predicted, only the vaccinated sheep lived. He later developed a rabies vaccine but could not identify the responsible organism. What he didn't know was that rabies was caused not by a bacterium but by a virus—a much smaller, as yet undiscovered, organism that reproduces only inside living cells.

If preventing disease was a rudimentary science until well into the 19th century, curing disease was no less so. The folk remedies that had been tried for centuries sometimes worked, but nobody knew why. A German bacteriologist named Paul Ehrlich first proposed the idea that infectious disease could be controlled or cured through the use of chemicals. Ehrlich's concept, which he called chemotherapy, was a radical new approach to treating illness. It was he more than anyone who designed the blueprints for modern medicine.

Ehrlich was a man of maddening contradictions. A mediocre student, he became an insatiable reader. He attacked problems by trying everything; not surprisingly, he was wrong far more often than not. "Never had any searcher coined so many utterly wrong theories," wrote one of his biographers. Ehrlich was also a fabled quarreler, not above hurling a book at assistants who failed to follow his logic; yet he could

One of a new breed of inventor, the genetic engineer, pores over a flask containing synthetic DNA. Only two centuries ago, people believed that curses and bad air caused infection and that the mixing of parents' blood governed inherited traits. Scientific sleuthing, however, revealed that germs cause infectious diseases, which led to the invention of preventive measures and cures. Within the last 40 years, microbiologists have cracked the code of life, the structure of the DNA molecule that controls every cell and determines genetic makeup. Today's medical inventors make artificial body parts and manipulate DNA to create new organisms and products for medicine and industry.

be generous, modest, and self effacing.

Like Pasteur, Ehrlich fell in love with chemistry as a youth. At Breslau University he examined the dyes used to stain bacteria to make them visible under the microscope. He realized that the dyes reacted chemically with the organisms, not only coloring but also killing them. Of the hundreds of dyes, Ehrlich thought, there must be at least one "magic bullet" that could kill bacteria without harming healthy cells.

In 1901 the notion of magic bullets summoned visions of medieval sorcery, and Ehrlich was tagged with the epithet Dr. Fantasy. He snorted at the title and went to work, trailing ashes and smoke from one of his ever present black cigars. And indeed he did find a dye, called trypan red, that in some cases killed the parasite trypanosome that caused sleeping sickness and other diseases.

Ehrlich knew that trypan red contained nitrogen, which resembled arsenic in some of its chemical properties. If arsenic were substituted for nitrogen, he reasoned, the new compound might be like trypan red but pack a much more powerful punch. Ehrlich and his staff tested every arsenic-based compound they could find on laboratory mice. Most of the substances were worthless as bacteria killers, but one, the 606th

Louis Pasteur peers into a flask holding the rabies-infected spinal cord of a rabbit. In the early 1880s, the French chemist discovered how to make a preventive vaccine for rabies from spinal cords harboring the virus. His vaccine saved the life of a nine-year-old boy mauled by a rabid dog in 1885. Soon bite victims streamed to Pasteur's Paris laboratory.

Thirty years earlier, this giant of 19th-century science had shown that "microscopic organisms abound in the surface of all objects." Some ferment sugar to alcohol or milk to cheese; others spoil food. Each microbe thrives in a distinct niche, Pasteur found. A page from his notebook (above) describes work on bacteria that grow without oxygen.

Robert Koch (third from left) and aides scrutinize blood samples in East Africa for the microbe that causes sleeping sickness. The bite of a tsetse fly (left) had infected the reclining man. By 1906, when the great German bacteriologist took this trip, he had expanded on Pasteur's work and identified the bacteria that cause anthrax, cholera, and tuberculosis. Scientists still use Koch's research methods.

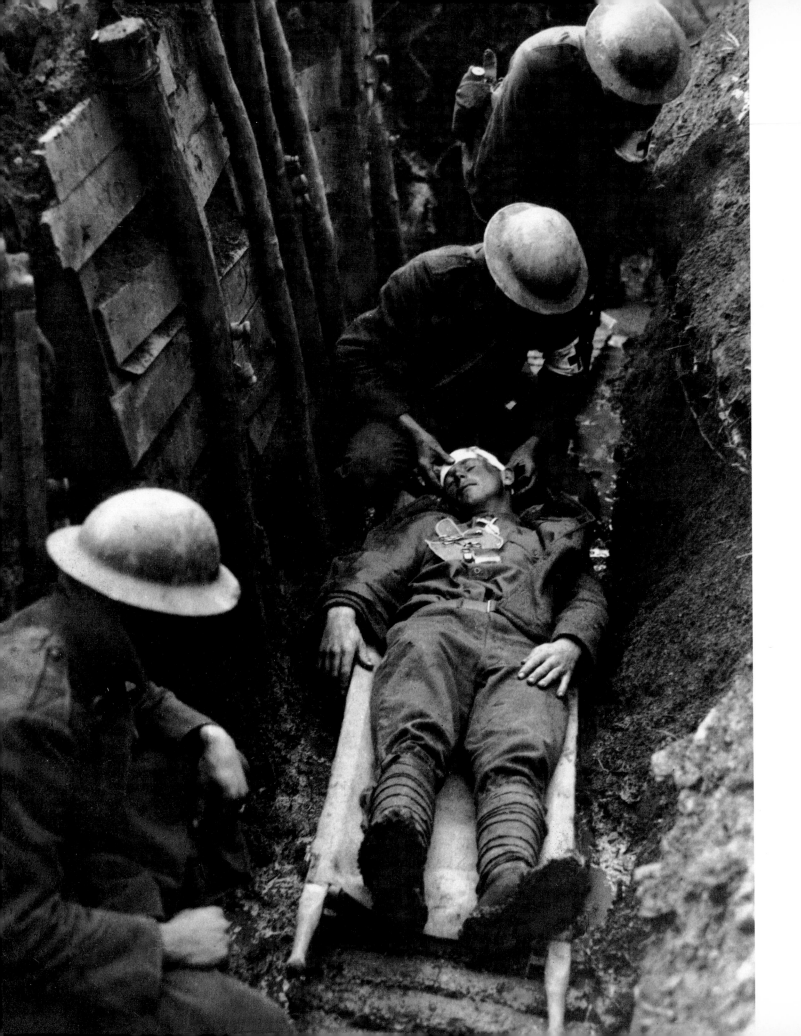

preparation, was different. Not only did it destroy trypanosomes, but it did so without the side effects that often accompanied trypan red.

Had Ehrlich stopped there, he would have been assured a respectful mention in medical history. In the course of his reading, however, he learned of a possible relationship between trypanosomes and spirochetes, the corkscrew-shaped organisms that cause syphilis. Ehrlich's tests, first in animals, then in humans, found in 1908 that indeed number 606 could cure syphilis.

Ehrlich accrued world honors for his invention of 606, or Salvarsan, its trade name. But even more significant were the thorough, painstaking methods he established for finding chemical bacteria-killers; these opened the way to the discovery of many other potent drugs.

Now, in a space of less than 75 years, medicine gained two formidable weapons against disease: the knowledge that germs cause infection and that these germs often succumb to specific chemicals. Although few realized it, the age of the wonder drug had dawned.

A fortuitous discovery gave us a powerful, naturally occurring drug to treat infections caused by bacteria. Penicillin might never have been identified at all had a reticent Scottish bacteriologist named Alexander Fleming been less observant and insightful one day in 1928 at St. Mary's Hospital, London.

Fleming's interest in antibiotics—drugs derived from microbes—began during World War I, when, as a captain in the medical corps, he saw men suffering and dying "without... anything to help them." Back in London, his research led him to the discovery of a protein in tears and mucus that killed laboratory bacteria. A few years later, in the course of experimenting with the bacterium staphylococcus, Fleming left several culture dishes exposed to the air. After awhile he spotted specks of mold in the dishes. He looked closer and saw that wherever

Penicillin, a "magic bullet" against infection, was born of the carnage of war. The death toll from infected wounds during World War I so distressed medical officer Alexander Fleming (above) that he vowed to find a drug that would destroy microbes without harming people. Some 13 years later, the Scottish bacteriologist spotted a banal sight: mold on a plate of bacteria. But around the mold the yellow clumps of bacteria had dissolved, leaving clear drops. Fleming named the microbe killer after the penicillium mold that secretes it. In 1939 Howard W. Florey and Ernst B. Chain at Oxford University devised a method for making penicillin practical. World War II saw an urgent campaign to mass-produce the drug for troops: A technician in England in the 1940s (above left) sterilizes flasks that will grow the mold.

there was a bit of mold, no bacteria grew—there were only clear drops that looked like dew. Not given to colorful speech, Fleming mumbled that he might have a mold "that can do something useful."

The mold turned out to be penicillium, similar to the greenish variety often found on stale bread. Fleming tried to extract pure penicillin from the mold but was unsuccessful. His discovery stirred little interest. Just before World War II, in fact, one American university would not donate a hundred dollars for purification studies. Despite Ehrlich's success with Salvarsan, many experts still doubted that organisms could be destroyed once they invaded the body. One of Fleming's mentors maintained that "drugs are a delusion."

Another reason for penicillin's slow acceptance was the difficulty of making it in large quantities; the mold often simply refused to grow in the laboratory, and the drug quickly broke down. But in 1940 two Oxford University chemists, Howard W. Florey and Ernst B. Chain, produced large volumes of pure powdered penicillin—more potent than the crude penicillin Fleming had found—with a freeze-dry method like that now used to make instant coffee. World War II prevented the mass production of penicillin in Great Britain, so drug companies in the United States took over manufacturing it on a large scale.

During the war, penicillin contributed to a 95 percent recovery rate of wounded Allied soldiers. By 1945, the year that Fleming, Florey, and Chain won the Nobel prize, the U. S. drug industry was producing 650 billion units of penicillin a month. Its development spurred the discovery of many other antibiotics that, in industrialized countries, slashed the death rate from bacterial infection.

Penicillin proved useless against one particularly virulent organism: the kind that causes tuberculosis, the lung disease known as the white plague. By 1940 TB was killing nearly 60,000 Americans a year and sending tens of thousands more to sanatoriums.

The man who changed that was a microbiologist named Selman A. Waksman, whose farm upbringing, first in tsarist Russia and later in New Jersey, gave him a special interest in soil-dwelling microorganisms. At Rutgers University he studied thousands of such organisms for their antibiotic properties. Finally, in 1943, his research assistant extracted a moldy clump of dirt from the throat of a chicken. The mold turned out to be streptomyces, and it killed the TB bacillus. Following several months of testing and purification, streptomycin was first used in 1944 to save a TB patient at Minnesota's Mayo Clinic. A decade later

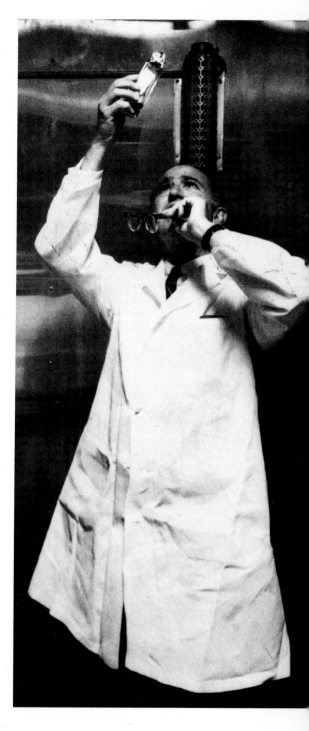

The iron lung—ominous symbol of poliomyelitis from the 1930s through the 1950s—forced air in and out of a patient's paralyzed lungs. In 1954 polio struck nearly 40,000 Americans and killed hundreds. The epidemics, usually during warm weather, turned summer into a time of terror. Fear of contagion cancelled vacations and quarantined entire communities. Hundreds of researchers labored four decades to conquer the three types of polio virus. American scientist Jonas Salk (right) was lionized in 1955 after creating a safe and effective polio vaccine derived from dead viruses. By 1961 his vaccine slashed polio cases in the United States by 96 percent.

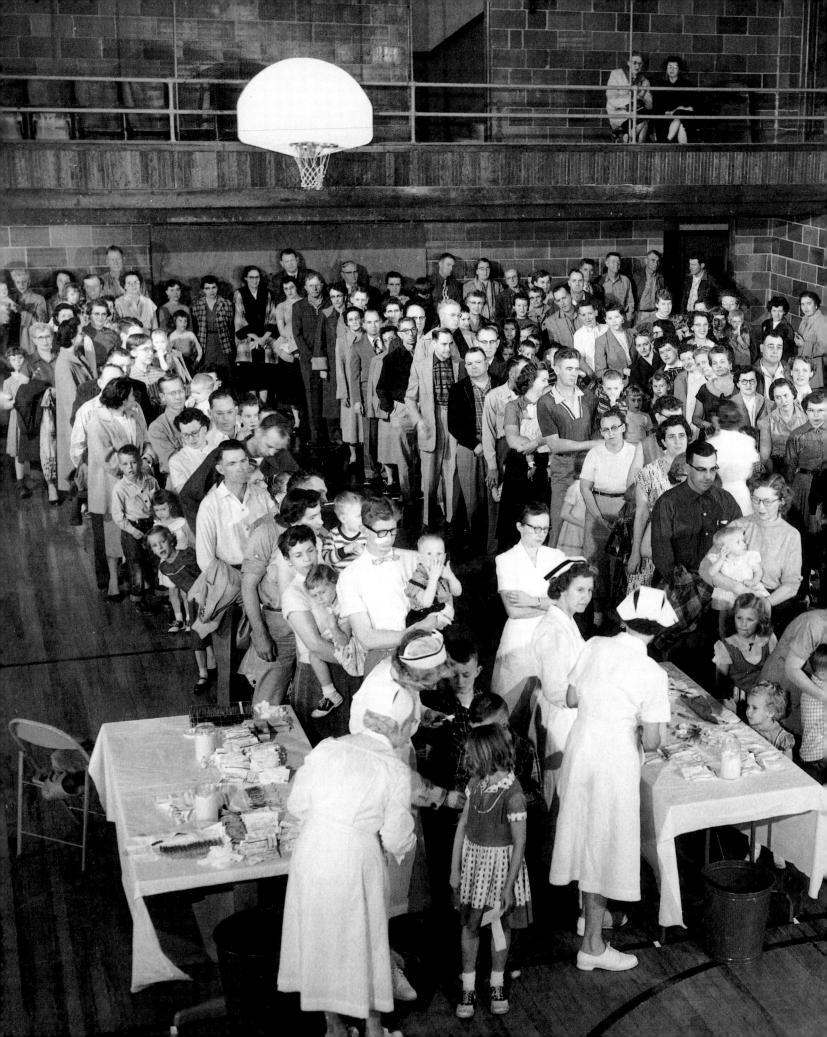

the annual number of TB deaths had dropped to 15,000. Today, 95 percent of TB cases are quickly tamed with a combination of streptomycin and a chemical called isoniazid, or another antibiotic, rifampin.

Not all diseases are caused by germs, of course. Some of the deadliest arise from an inborn defect. Others are due to a sort of internal mutiny, in which the immune system attacks the very cells it is intended to defend, causing an autoimmune disease.

One such disease is diabetes, the body's inability to absorb and metabolize sugars and other carbohydrates for use as fuel. Today one million diabetics in the U. S. control their disease with injections of the hormone insulin. Eighteen centuries ago, however, the Greek physician Aretaeus called diabetes "a melting down of the flesh and limbs" that made life "disgusting and painful." And as recently as 1921, it doomed the patient to death as essential nutrients were excreted in the urine. The two men who finally offered a reprieve—Frederick G. Banting and Charles H. Best—brought little experience to their task.

The son of a Canadian farmer, the shy and studious Banting trained briefly for the ministry but later became an orthopedic surgeon. With little demand for his services, however, he spent several hours a day reading medical journals. In late 1920 he read a report suggesting that diabetes might be cured with a substance produced in the pancreas, one of the major endocrine glands.

The theory wasn't novel, but it was controversial. Experimenters knew that the pancreas not only supplied digestive enzymes but also played a role in diabetes; around the turn of the 18th century, for example, a researcher named Johann Brunner made dogs diabetic by removing their pancreases. Scientists also knew that the pancreas contained certain unique cells (now called beta cells) that might produce the mysterious substance. But many experts remained unconvinced that such a substance even existed.

In the spring of 1921, the 29-year-old Banting asked a University of Toronto professor, J.J.R. Macleod, for permission to conduct diabetes studies. Macleod doubted that the young physician could make any headway against a disease that had long baffled medical experts; nevertheless, he gave Banting ten dogs and eight weeks' use of an unused laboratory—a dark, ill-equipped room that had to be scrubbed from top to bottom. He also threw in an assistant: the 21-year-old Best, a medical student skilled in deciphering the complex chemistry of the blood.

The two young men took to each other immediately. They were alike

Young residents of Protection, Kansas, bare arms for injections of Salk polio vaccine in 1957. At the time, Albert Sabin of Cincinnati was field-testing an oral vaccine based on live, weakened viruses. A soaked sugar cube (right) sufficed to confer immunity. The oral vaccine soon edged out Salk's vaccine in the U. S., though not in Europe. Both vaccines are effective, and newer ones are being developed.

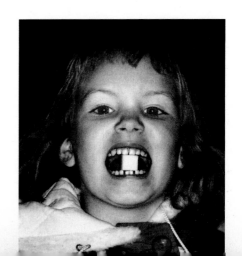

in temperament and upbringing, and both had witnessed the horrors of diabetes. Best had lost a favorite aunt to the disease; Banting had watched a neighbor girl succumb just a few months after her diagnosis, her bed wrapped in the fatal sweet fragrance that signaled an excess of unmetabolized sugar in the blood and urine.

Their studies got off to a discouraging start. Most of their two-month allotment was used up by failures, and the brutally hot summer of 1921 made it almost impossible to control the odors from the dogs' cages.

Then Banting had an idea: Might the sought-after substance (which Banting now called isletin) be destroyed along with the pancreas's digestive enzymes when the organ was mashed up? That might explain why experiments in which material from mashed pancreases was injected into diabetic dogs showed no results. To find out, the young researchers securely tied off the dogs' pancreatic ducts; as in earlier experiments, the organs degenerated. Banting speculated that while the decay destroyed the enzymes, the beta cells—and the precious isletin—would be left intact.

He was right. That summer the men injected a diabetic dog with isletin, then watched in amazement as the feeble animal bounded back to health. True, the effects of the injections were short lived; the dog had to be dosed at least once a day, and pancreases were in short supply. But now the time had come to test the substance on a human patient.

The men found a volunteer, a thirteen-year-old boy named Leonard Thompson with a two-year history of diabetes. To minimize the deadly sugar levels, doctors had kept the boy on a daily diet of only 450 calories. Bedridden, he weighed a mere 65 pounds. The isletin, which later became known as insulin, made possible a spectacular recovery.

Discoveries of other hormones followed quickly over the next few decades, notably cortisone and the sex hormones estrogen and progesterone. Today cortisone relieves millions who suffer from rheumatoid arthritis and other diseases. In 1960 progesterone appeared on the market in a tiny tablet that would dramatically affect Western economics and human behavior: the Pill.

The search for new drugs goes on continuously. In 1969, while Swiss researchers were analyzing soil fungi for antibiotics, they discovered a chemical in a Norwegian fungus. Three years later, Swiss immunologist Jean-François Borel found that this chemical, named cyclosporine, suppressed the part of the immune system that rejects foreign tissue. The drug has enabled thousands of successful kidney, liver, heart, and pancreas transplants. In 1974, 80 percent of heart transplant patients died within the year. A decade later, after cyclosporine became available, 80 percent survived at least two years.

Physicians had dreamed of transplanting organs since ancient times. A kidney was successfully implanted in a human in 1950, but it was the first human heart transplant that captured the world's imagination. Christiaan Barnard performed the surgery at the Groote Schuur Hospital in Cape Town, South Africa, in December 1967. The donor heart, from a 25-year-old auto accident victim, worked beautifully, but the patient died 18 days later of a lung infection. Barnard was devastated: "I

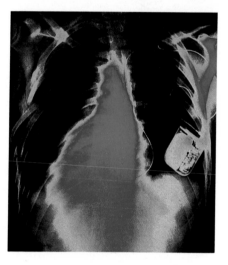

To keep hearts beating steadily, bioengineers invented the pacemaker (above) in the 1950s. Heart transplants first succeeded in 1967, permitting one patient to ponder his own diseased organ as a donor's heart beat inside his chest (opposite). Pioneered by Christiaan Barnard, Michael DeBakey, and Adrian Kantrowitz (top), heart transplants proliferated after the licensing of cyclosporine in 1983, a drug that lowers the body's ability to reject foreign tissue. Today over a hundred U. S. medical centers perform the surgery on patients whose numbers are limited by the 1,000 to 2,000 donor hearts available yearly.

Willem J. Kolff

"Even without hope you should persevere," asserts Willem Johan Kolff, pioneer in the field of artificial organs. Born in Leiden, the Netherlands, in 1911, he learned early about tenacity and the quest to heal. Kolff's father directed a tuberculosis sanatorium when much of the treatment was rest, good food, and fresh air. On long walks in the woods, the elder Kolff would confide in Willem his feelings of helplessness as he watched a patient dying.

His father's frustration became his own when, as a 27-year-old physician working toward his Ph.D. at the University of Groningen, Willem Kolff cared for his first patient dying of kidney failure. The young farmer had lost his sight and vomited repeatedly as wastes—normally excreted by the kidneys and eliminated in urine—accumulated in his blood. To Kolff fell the task of telling the man's elderly mother that her son was about to die. If only, Kolff thought, I could remove 20 grams of the waste product urea from the patient's body each day, the man might live.

Scientists before Kolff had dreamed of building a machine to cleanse blood outside the body, but no one had yet succeeded. Then, in 1938, a biochemistry professor introduced Kolff to "the wonders of cellophane." Kolff filled a cellophane sausage casing with blood and urea and swished the tubing in salt water. Within five minutes all the urea passed through the cellophane membrane into the water, while the blood remained inside the casing: The blood had been purified, or dialyzed.

In 1940 German troops invaded the Netherlands. Kolff set up the first blood bank in Europe, at The Hague. Then, with his wife, Janke, and the first of their five children, he left Groningen when the occupying forces

Bionics pioneer Willem Kolff (top). His 1940s dialysis machines—the world's first—await shipment (above).

installed a Nazi to run the university's medical department.

Kolff found work at a hospital in the old port city of Kampen. There, in 1943, he built a "rotating drum" artificial kidney, a device comprised of 65 feet of sausage casing wound around a horizontal aluminum drum that was partially immersed in salt water. Kolff's assistant would draw blood from a patient into a flask, hoist it to the ceiling with a pulley, then let the blood drain through the tubing as a motor turned the drum slowly in the saline bath. During the war, the physician made seven more dialysis machines out of wood since aluminum was scarce, scrounging the materials and scattering the devices throughout town to

protect them from bombs.

Though patient after patient died, Kolff persevered, continually tinkering with the technique. Finally, in 1945, he saved the life of a 67-year-old woman, dialysis patient number 17.

Eager to share his invention with the world, Kolff shipped his machines, unpatented, to hospitals in London, New York, and Montreal. After emigrating to the United States and joining the Cleveland Clinic in 1950, Kolff adapted a Maytag washing machine for home dialysis. Later, he wound cellophane tubing and window screening around seven juice cans and created the prototype of a disposable dialyzer. Elaborating on experiments in the Netherlands, Kolff also developed an early heart-lung machine.

In 1957, with Tetsuzo Akutsu, Kolff created the first artificial heart in the United States and implanted the plastic device into a dog. "The medical community wanted nothing to do with such a thing," he says. "The idea that the symbol of love...might be replaced by a mechanical pump was very difficult for people to accept."

The self-proclaimed "oldest artificial organist" has served for years as an influential mentor. In 1967 he founded the Institute for Biomedical Engineering at the University of Utah; over the years he trained so many researchers of diverse "spare parts" that Salt Lake City became dubbed Bionic Valley. There, in 1982, the first human, Barney Clark, was fitted with an artificial heart. Today such hearts serve as bridges to donor-heart transplants.

Now retired as director of the institute, Kolff still trains researchers and works on an artificial heart that can be mass-produced. He remains optimistic about bionics. Because artificial hearts will not tire, they may someday prove more effective than their natural counterparts, he says: "I predict that the winner of the Olympic marathon of the year 2000 will be disqualified for wearing an artificial heart."

Carole Douglis

Portable machines (opposite) permit kidney patients to dialyze while camping.

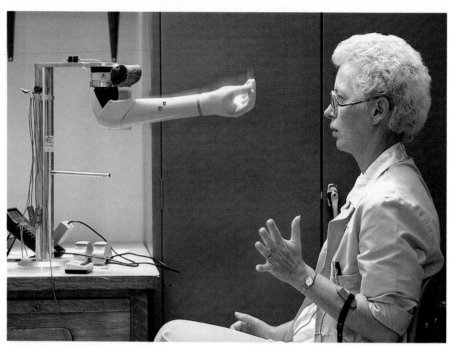

crash-landed," he said later. Yet he persisted, applying what he had learned to a second patient. The man, who had been bedridden, was soon able to get up, drive, and even swim in the ocean. He lived more than 18 months.

The success of transplant surgery since cyclosporine has created a new challenge perhaps as thorny as the technical ones already overcome: the shortage of donor organs. Each year some 12,000 Americans die of diseases that heart transplants alone might cure.

Researchers in bionics—the science of designing devices to replicate body functions—hope someday to offer artificial parts as an acceptable substitute. The attempt to build spare parts for humans is ages old. Early physicians fashioned peg legs for amputees, and everyone knows about George Washington's ill-fitting ivory teeth. But in this century inventors launched a more audacious quest: to replace joints and organs.

Already two to three million Americans a year receive a cardiac pacemaker, artificial hip, electric limb, Dacron blood vessel, finger- or toe-joint replacement, plastic lens implant in the eye, or other bionic part. Hundreds of thousands of others are alive today thanks to artificial organs that are not implanted: the kidney dialysis machine, for instance, and the heart-lung machine, the two oldest types of artificial organs.

In 1931 Boston surgeon John H. Gibbon watched a woman die on the operating table while her circulation was stopped for six and a half minutes, the time it took her doctor to cut a blood clot out of her lung. "I kept thinking that we could have saved that poor woman's life," Gibbon said later, "if we could just have taken some blue blood out of her veins, put oxygen in, and let the carbon dioxide escape."

Gibbon and his wife spent the next 20 years trying to build a machine that would do just that. He hooked up one of his contraptions to stray cats captured at night on Beacon Hill with the help of canned tuna fish. The result of their work—the first machine that could take over the function of heart and lungs during operations on those organs—has been hailed by one medical historian as "among the boldest and most successful feats of man's mind." In 1953 the machine permitted successful heart surgery on an 18-year-old girl. But after two other patients died, Gibbon grew discouraged and took his device back to the lab. In the meantime, cardiologist C. Walton Lillehei developed a simpler, more reliable heart-lung replacement. Its descendants enable U. S. surgeons to perform some 100,000 open-heart operations a year.

Lillehei also took part in creating an electrical bionic part that helps keep the beat in millions of American hearts. Today's pillbox-size pacemaker runs on a battery and metes out tiny, painless electrical signals to regulate the heart's pumping. But its precursor, a stimulator developed in 1952 by Paul M. Zoll at Harvard Medical School, walloped up to 150 volts of electricity to the heart through electrodes on the chest. "The pain was intolerable," said Lillehei. "You just couldn't keep those electrodes on, not even if you tied them down physically. They burnt the skin." In 1957 Lillehei hooked wires directly to a patient's heart. Just two volts could jolt a stopped heart into action, but the patient remained wired to a large machine. A few months later Lillehei recruited

The body's own electrical signals control the state-of-the-art Utah Artificial Arm and a hand that can attach to it (opposite, lower). Electrodes inside the arm pick up impulses from the stump muscles; amplified and directed by electronic circuitry (opposite, upper left), these signals tell the elbow to flex and straighten, the wrist to rotate, the hand to open and close.

A technician puts the arm-hand unit through its paces during a five-hour final test (opposite, upper right). Made in West Germany, the hand has greater grip strength than a man's. The wrist rotates 360 degrees and, with elbow locked, the arm can carry heavy loads. Says Stephen C. Jacobsen (below), whose team designed the arm at the University of Utah, "Our objective is to get machines to move with the grace, strength, and precision that people have."

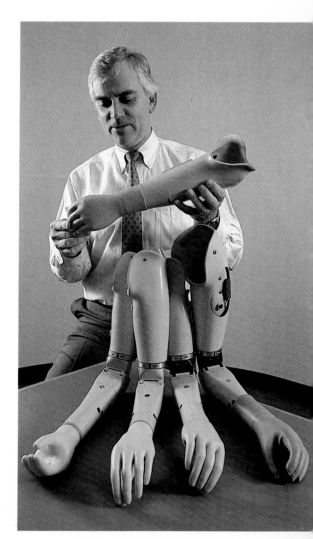

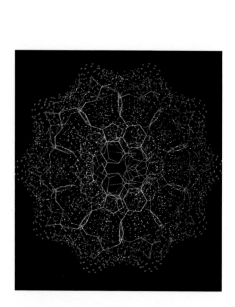

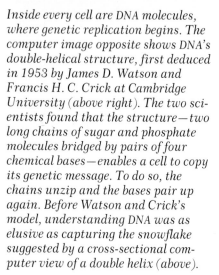

Inside every cell are DNA molecules, where genetic replication begins. The computer image opposite shows DNA's double-helical structure, first deduced in 1953 by James D. Watson and Francis H. C. Crick at Cambridge University (above right). The two scientists found that the structure—two long chains of sugar and phosphate molecules bridged by pairs of four chemical bases—enables a cell to copy its genetic message. To do so, the chains unzip and the bases pair up again. Before Watson and Crick's model, understanding DNA was as elusive as capturing the snowflake suggested by a cross-sectional computer view of a double helix (above).

an electronics repairman named Earl Bakken to devise a portable power source: a small, transistorized box that could be hung on a patient's belt. Soon, even the external wires disappeared from the chest as researchers found a way to implant pacemakers.

But few bionic ventures are as dramatic and controversial as the attempt to replace the heart itself. To craft a successful heart has been the grail of many bioengineers since 1937, when a Russian scientist replaced the hearts of dogs with mechanical pumps. The devices kept some of the dogs alive up to five and a half days. The best known of the recently designed hearts is the Jarvik-7, named after Robert K. Jarvik, a protégé of the prodigious Willem J. Kolff at the University of Utah, and one of 247 researchers who helped work out the complicated design.

The plastic, grapefruit-size Jarvik-7 is driven by rhythmic pulsings of compressed air that a several-hundred-pound machine sends into the heart through plastic tubes in the chest. Because recipients face the threat of stroke and infection and must remain tethered to the power supply, such hearts are now used almost exclusively as temporary bridges to heart transplants. However, prototypes are on trial for totally implantable hearts run by electricity.

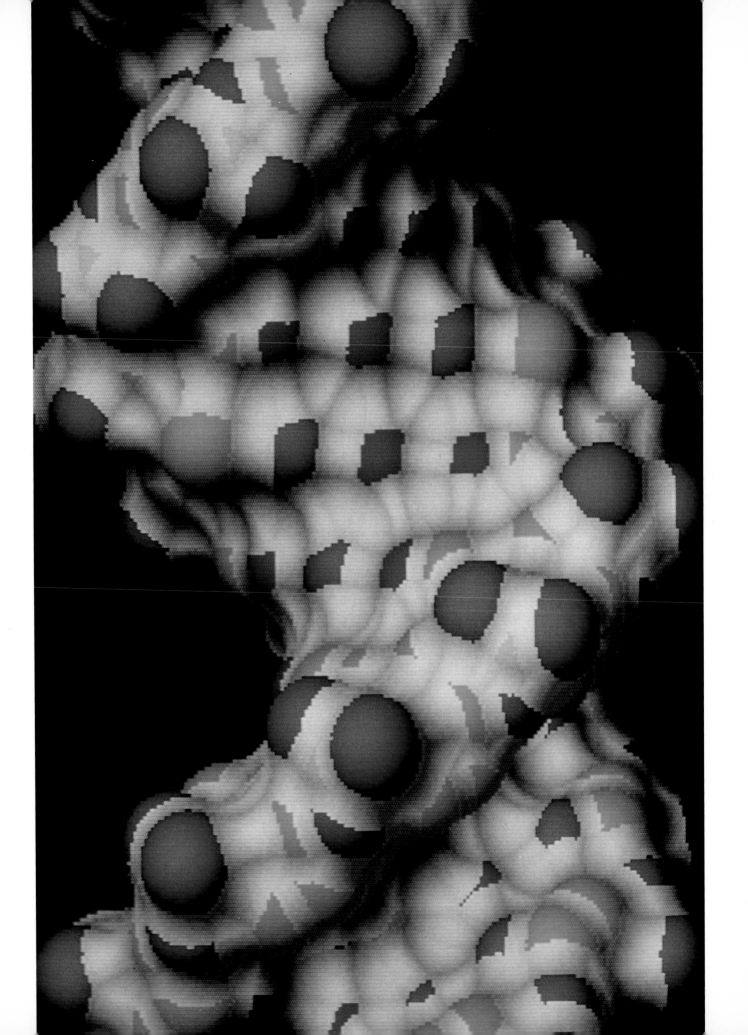

294

Bioengineers work to mimic the function of other organs too: skin, blood, even cells. Imitation skin, developed by Ioannis V. Yannas at the Massachusetts Institute of Technology and John F. Burke at Massachusetts General Hospital, spares third-degree burn victims much of the infection and scarring that accompany their wounds. Bionics scientists agree that, in the future, not only skin but bionic parts such as joints and tendons, bones, and organs may be temporary devices. Made from resorptive materials and impregnated with human cells, they may be used as templates to regenerate new tissues, organs, and limbs.

The body's workings are so stunningly complex that duplicating some of them may stymie medical inventors for decades—or forever. In the meantime, though, biologists are learning to interpret the very core of life, and in so doing, compete with nature itself in composing new life forms. Once again, this adventure began in the 19th century.

While Pasteur was proving that germs can cause disease, Gregor Mendel, an Austrian monk, was demonstrating that invisible *elementen* govern basic laws of heredity. He bred 28,000 pea plants in his monastery garden during a ten-year experiment. Mendel's work showed that traits such as tall or short stalks, white or purple blossoms are inherited from one generation to the next in predictable ratios. Unlike Pasteur's findings, Mendel's were completely ignored during his lifetime; an appointment as abbot was his sole earthly reward.

Biologists would have to work nearly another century before identifying the active ingredient in Mendel's elementen. In 1944 Oswald Avery and his colleagues at New York's Rockefeller University proved that the molecule deoxyribonucleic acid (DNA) determined the traits of living things. The scientists showed that DNA underscored the common ancestry of all life, whether a microbe, a water flea, or a human.

But what did the DNA molecule look like, and how did it work? It appeared too simple to account for the innumerable characteristics of all the world's animals and plants—just long chains of sugar molecules alternating with phosphates and four types of chemicals called bases.

Enter James D. Watson, an American biologist, and Francis H. C. Crick, a British biophysicist. The two spent 18 months in a Cambridge, England laboratory working with Tinkertoy-like molecular models to try to deduce how DNA might be configured. They based some of their hunches on X-ray pictures of crystallized DNA that seemed to show a double chain of some sort. Finally, in 1953, Watson and Crick announced the solution to one of nature's great puzzles.

The "most golden of all molecules," as Watson called DNA, turned out to resemble a long, twisted ladder—a double helix. The sugars and phosphates form the ladder's sides. The ladder's steps consist of pairs of bases. Thousands of the base pairs make up a single gene. The sequence of the bases within the gene determines the type of protein a cell is to make—whether a sex hormone or fingernails—and slight variations in those sequences decide an individual's unique genetic inheritance. The blueprint for a human life is encoded in about 100,000 genes, grouped into 46 rodlike bundles called chromosomes. If uncoiled, the DNA wound in the chromosomes of a single human cell

California biochemist Herbert Boyer scrutinizes a vial of recombinant DNA. With Stanley Cohen in 1973 he pioneered genetic engineering—the splicing together of DNA from different organisms to create new genes that alter nature.

Treatment with synthetic human growth-hormones helped Tracy Moreno grow five inches in one year (opposite, upper). A human growth-hormone gene introduced into a mouse embryo doubled the size of a rodent (opposite, lower). Commercial livestock breeders hope that applying this technique to farm animals will produce superanimals—fast-growing pigs and cows that resist disease. In April 1988 the U. S. Patent and Trademark Office granted a patent for a genetically altered mouse, the first patent ever issued for an animal.

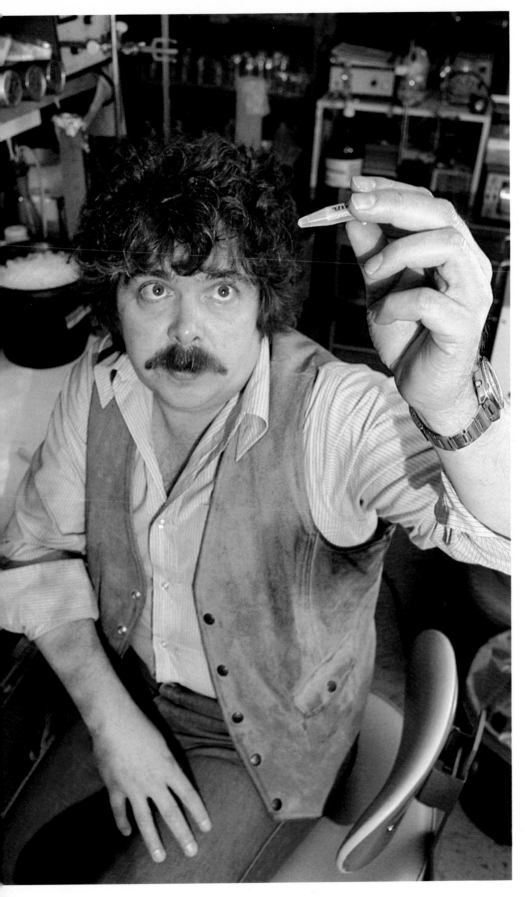

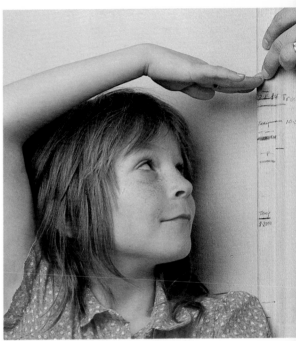

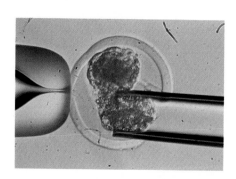

A glowing tobacco plant highlights a powerful new genetic-engineering tool (opposite). By inserting the luciferase gene, which makes fireflies luminous, into a plant's DNA along with an introduced trait, biologists can see where the plant has employed that trait. When linked to a wound-response gene, luciferase stimulates an injured plant to shine.

Bioengineers can also divide embryos. Inside a gelatinous sac (above), a microsurgical blade has split a week-old cattle embryo in two. Tweezers transfer the embryos to cows that will carry both to term. Such manipulation yields genetically identical twin calves (above right) and increases reproduction rates.

would stretch 6 feet and contain information equivalent to 600,000 printed pages.

The discovery left much personal bitterness in its wake; one researcher later wrote that "DNA . . . is like Midas' gold. Everyone who touches it goes mad." Understanding the molecule's structure unleashed tantalizing possibilities for changing it, and therefore manipulating the foundation of life itself.

The most fateful experiment of the gene age was planned in a Honolulu deli in 1972. Stanley Cohen, a doctor from Stanford University, and Herbert Boyer, a University of California biochemist, in Hawaii for a meeting, speculated about what might happen if a particular gene was cut from one species and inserted into another. Earlier in the year Paul Berg at Stanford had combined the DNA of two different types of virus, creating the first recombinant DNA molecule. Now, back in his lab, Cohen snipped a single gene from a toad cell—like removing a pearl from a microscopic necklace—and transferred it to the DNA of a bacterium. As the bacterium reproduced, the gene found its way into all the bacterium's descendants. The gene had been cloned; the age of genetic engineering had begun.

Whereas Ehrlich created drugs from chemicals and Florey and Chain from mold, the genetic engineers have mass-produced substances made by the human body itself. "The single most profound and . . . most significant advance in biological science in the 20th century," physician Lewis Thomas calls the technology of taking apart and recombining DNA to make new organisms. So far, it has given birth to only a handful of drugs, but hundreds of companies around the world are staking billions of dollars on the pharmaceutical and agricultural products they expect to create. Among them is a supervaccine, made of "cassettes" of strung-together genes, that can protect humans against smallpox, hepatitis B, the flu, herpes, meningitis, and maybe another half-dozen illnesses with just one shot.

Biotechnology promises crops with genes that make them resist diseases and pests. Already botanists have cut out a bacterial gene that kills a plant virus and one that kills insect pests, and inserted them into tomato plants. They have altered nitrogen-fixing bacteria to use atmospheric nitrogen more efficiently, so that farmers could save money on fertilizer. They have engineered bacteria that prevent frost crystals from forming on strawberries and potatoes and ruining the crops.

Not surprisingly, critics have accused genetic engineers of playing God. Even some scientists balked amid the heady successes of early biotech. To quell fears about the dangers of genetic recombination—the "gruesome parade of horribles" one observer envisioned—the federal government has created extensive regulations to govern the release of man-made organisms. Biotechnologists worry that the tangle of regulations could strangle the industry.

Although, as a geneticist noted, "the scenario of a giant rutabaga crawling out of the lab and devouring Cleveland was overstated," many critics are not appeased. Jeremy Rifkin, among the most vocal, feels that by changing the genetic structure of life "we have the potential to do irreparable psychological, environmental, moral, and social harm to ourselves and our world."

But the perils of life engineering must not blind us to its promises. It may give us the tools to prevent and cure many of our most feared diseases, feed the world's hungry, and begin to understand the nature of life. Ethical dilemmas have been a part of medicine since long before Jenner's day. What has changed is the breakneck pace of medical knowledge, which today is believed to double every five years. As it increases, so does our power over the world for both good and ill.

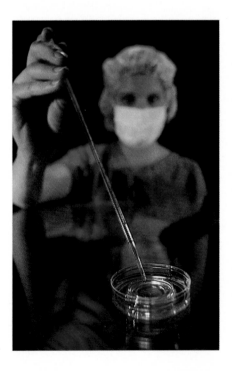

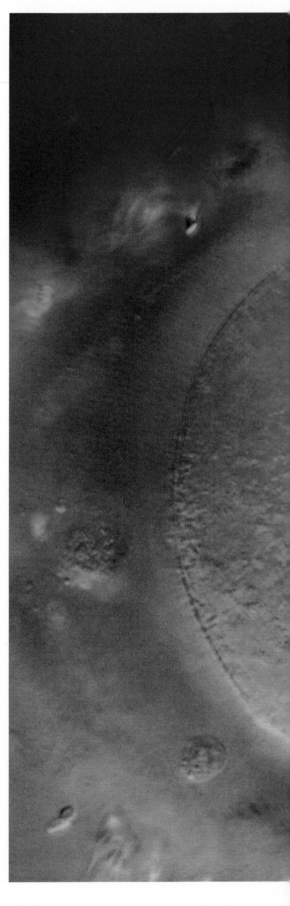

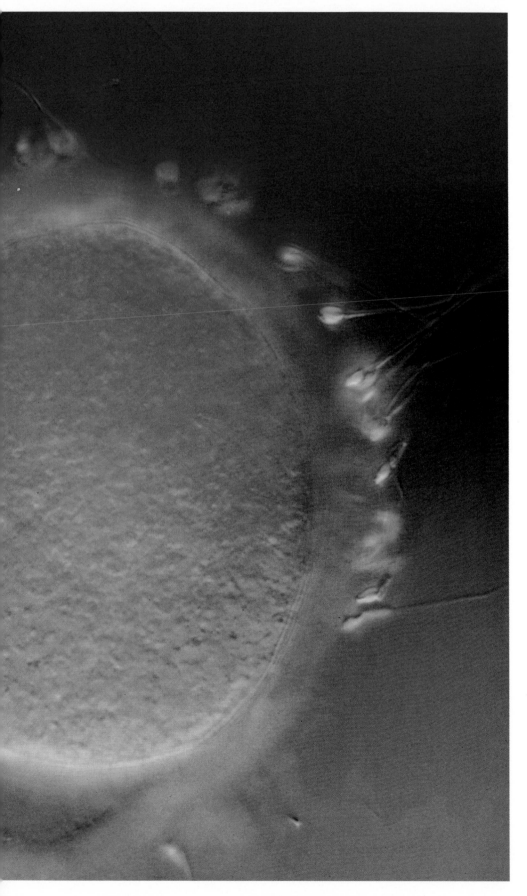

Only one sperm need penetrate a human egg to fertilize it (center). A precisely timed biological routine, this meeting in a fallopian tube usually generates a new life. But when blockage or disease prevents normal conception, technology can sometimes outwit nature. In a laboratory dish (opposite), a man's sperm join a woman's eggs in the procedure known as in vitro fertilization. Up to 72 hours later, once hormones have prepared the uterine lining, the developing embryo is inserted into the patient; yet successful delivery follows less than a fifth of the time. Since the 1981 birth of Elizabeth Carr (below), the first test-tube baby in the U. S., bioengineers have developed embryo transfer and frozen preservation of embryos as weapons against infertility.

Technology of Tomorrow

By Carole Douglis

Tomorrow, discoverers and inventors yet unborn will create a world of new materials, intelligent machines, and medicines that we can scarcely imagine. Today, in posh corporations and basements, in universities and government laboratories, technological geniuses are setting the stage.

In 1983 a self-professed "greenhorn and outsider" started exploring an arcane corner of physics. At IBM's Swiss laboratory, K. Alex Müller, a courtly physics professor, and his German protégé, J. Georg Bednorz, searched for a cheap, simple-to-make superconductor, a material that conducts electricity perfectly, without losing any energy to resistance. Guided by an observation and a hunch, they were screening compounds that typically fail to conduct an electric current at all. Bednorz recalls his fear that other scientists would say, "These guys are crazy!"

The phenomenon of superconductivity has tantalized scientists since Dutch physicist H. Kamerlingh Onnes discovered and named it in 1911. In everyday wires the speeding electrons that comprise an electric current waste energy ricocheting against other atoms and each other, generating heat, or resistance. But at colder temperatures electrons collide less often. Chilled almost to absolute zero—minus 460°F, the theoretical temperature where all atomic motion would stop—the electrons of a superconductor glide freely, with no resistance whatsoever. Another peculiarity of superconductors: They repel a magnetic field, whether or not they are conducting electricity. Place a magnet above a superconductor, and it floats in midair. Place a superconductor above a magnet, and it too will levitate.

Scientists labored six decades to coax superconductors to work at more hospitable temperatures. Yet by 1973 they had merely raised the transition temperature—that below which superconductivity switches on—to a frigid minus 418°F. There it stuck for 13 years. Funding sources dried up, experts drifted into other specialties. Cooling superconductors to such temperatures was expensive and difficult, requiring heavy insulation. It limited their commercial use to large devices like magnetic resonance imaging machines, which help doctors diagnose disease by measuring the body's reaction to magnetic fields.

Bednorz used a mortar and pestle to blend, and an oven to fire, 50 different ceramic compounds. But after two years he and Müller were stumped. Then Bednorz noticed an article in a French journal about an unusual material—a combination of copper and oxygen, laced with barium and the rare earth element lanthanum. Bednorz tried to reproduce this copper oxide in order to test it for superconducting properties, but by "a kick of luck," he set his oven temperature low and accidentally prepared a slightly different version. His recipe superconducted at a record-breaking minus 406°F. Bednorz had revealed a whole new class

K. Alex Müller and J. Georg Bednorz will deserve much of the credit if we enter an age of superconductors— materials that allow electricity to flow without resistance and power loss. Since 1911, scientists have observed superconductivity at temperatures near absolute zero, but not until 1986, in IBM's Swiss laboratory, was a material found that would work under warmer conditions. It won Müller and Bednorz a 1987 Nobel prize. Now the race is on to discover a practical, high-temperature superconductor.

Superconductors repel magnetic fields: They float over magnets and vice versa. Scientists dream of the day when superconducting trains will fly inches above tracks at 300 miles an hour, surpassing the speed of a West German model (opposite), which levitates by means of conventional magnets.

of superconducting materials.

Soon teams headed by Ching-Wu "Paul" Chu at the University of Houston and Maw-Kuen Wu at the University of Alabama broke another barrier by fabricating a compound that worked at minus 283°F. Scientists could now use liquid nitrogen—cheaper than milk and as abundant as air—to invoke superconductivity. The discovery triggered an epidemic of "Chu's disease," said one physicist: a round-the-clock, round-the-world chase to reproduce the material that Chu's team had made—a copper oxide mixed with barium and yttrium. That temperature limit has since been bettered with newer ceramics, and scientists have fleetingly glimpsed superconductivity at room temperature.

"This could well be the breakthrough of the 1980s," says Alan Schreisheim, director of the Argonne National Laboratory near Chicago, "in the sense that the transistor was the breakthrough of the '50s." Making superconductors practical is the prize that now beckons hundreds of laboratories worldwide. Today's high-temperature superconductors lose their special properties when carrying a heavy load of electricity, and making the brittle materials flexible is like "turning a coffee cup into a wire," says IBM researcher Paul Grant.

Among the payoffs closest to realization may be SQUIDS—superconducting quantum interference devices—which measure infinitesimal amounts of magnetism and electricity. SQUIDS can help diagnose brain function without surgery by sensing the electrical firing of brain cells from outside the head. They can also help locate minerals, oil, and water. The further future may find superconducting electric motors and appliances that increase in power as they shrink in size, giant underground electromagnets that store electricity overnight for use during peak hours, and superconducting cables that save billions of dollars in electric bills in the industrialized West and help bring cheap electricity to villages of the Third World. Transportation officials envision a network of sleek trains flying above the ground on a magnetic cushion, competing with airplanes on crowded interurban routes. Physicists daydream of floating toys, cars, furniture, people. "You could put on a pair of special shoes," says Praveen Chaudhari of IBM. Floating above superconducting tracks, "you could push yourself and keep going."

Superconductivity could enable computer designers to pack more electrical circuits onto computer chips, and more chips into a computer, without the danger of heat from resistance damaging delicate components. Especially interested in ever smaller and more powerful mechanical brains are researchers of machine, or artificial, intelligence. Their goals: to better understand human intelligence and create machines that think, see, move, and talk like people.

Some 30 years ago, pioneers in artificial intelligence created programs that seemed astounding. One beat human champions at chess, another proved mathematical theorems, a third got an A on an MIT final exam in calculus. A brilliant humanoid, it seemed, was just around the corner. But prospects changed when programmers tried to instruct machines to do things four-year-olds master easily. As a graduate student at Stanford University, Hans Moravec spent six years teaching a

Physicist Ru-Ling Meng shows a student the mold in which she will pack the ingredients—now a fine powder—to make a superconductor. This University of Houston lab has concocted and tested tens of thousands of materials under the leadership of Ching-Wu "Paul" Chu. One of these made scientific history: A ceramic combining copper oxide, barium, and yttrium superconducted when cooled only to minus 283°F, topping Müller and Bednorz's breakthrough by 123 degrees.

A simple recipe for a superconductor (opposite, top to bottom): Grind the raw materials with a mortar and pestle. Bake in a crucible up to eight hours at about 1650°F. Then show it off: Chilled with inexpensive liquid nitrogen, a small magnet hovers in midair above a superconducting disk.

No one knows just how superconductors work. It appears that electrons, instead of colliding and wasting energy as they do in normal wires, move in pairs like surfers riding side by side on ocean waves. However, "superconductivity may turn out to have as many causes as the common cold," says Nobel physicist Robert Schrieffer.

robot—this one a wheeled cart with a TV camera for an eye—to see well enough to avoid bumping into things. The famous Stanford Cart took five hours to cross a room, Moravec recalls, pausing "to think for about 15 minutes" before rolling a few feet. Getting computers to use human language proved no simpler. "You have to learn chess, so there's some hope that . . . you could . . . explain it to a computer," says computer scientist Tomas Lozano-Perez, of MIT. "The stuff that's hard is common sense, vision, moving your arms. We don't know how we do it."

Today's inventors are both capitalizing on the idiot savant abilities of computers and trying to teach a contraption of silicon and metal to understand language, use a modicum of common sense, and find its way in the world. They have created sensors for smell, sound, touch. Some robots have a limited ability to recognize objects by vision, and can navigate using sonar and laser scanning as well as visible light.

It is "better to be knowledgeable than smart," decided Edward Feigenbaum of Stanford. In the 1960s Feigenbaum and colleagues designed the first expert system: a computer program, packed with knowledge gleaned from human experts, that can draw inferences and give advice in a narrow field of expertise. Expert systems reason in a chain of if-then statements: If the cheese has green mold on it, then throw it out. Inspired by Feigenbaum's success, Edward Shortliffe, a Stanford medical student and computer buff, tapped the reasoning several Stanford physicians used when treating patients with infections, and helped devise the expert system Mycin in the early 1970s. Mycin prescribed the right antibiotic more often than doctors did.

Thousands of expert systems now take on tasks from war games to computer-circuit design to financial analysis. Herbert Schorr of IBM says of the computer revolution: "The first wave automated data processing; the second will automate decision making." On the horizon lie expert systems with broader and broader expertise.

Yet specialists built of silicon still do not know what a human is, let alone the pain of illness or the motives for war. Douglas B. Lenat seeks to remedy such ignorance with a massive project called Cyc, a play on "psych" and "encyclopedia." Lenat's 35-person team—at Microelectronics and Computer Technology Corporation in Austin, Texas—is hand-coding a 30,000-entry encyclopedia into computer language, sentence by sentence. For each entry they explain assumptions a human reader would possess, including common sense concepts of time and space, emotions and motivations. Cyc should someday be smart enough to continue learning on its own by drawing connections between old and new information. Delivery date: late 1990s.

Bestowing such intellect on robots is a long-term goal of many in artificial intelligence. But innovators have taught even simpleminded machines to provide startling services. Yik San Kwoh, an electrical engineer at Long Beach Memorial Hospital, saw that brain surgeons cut through brain tissue to a tumor blind, or "freehand." He bought a robotic arm and created a special guide-tip that could aim a biopsy needle. Hooked up to a CT scanner—computerized imaging equipment— the robot points a needle toward the tumor with a tolerance of 1/2000 of

306

Piloted by computer, the experimental Autonomous Land Vehicle (ALV) scouts driverless through the foothills near Denver. Scanning for obstacles and pathways with video cameras and laser beams (simulated here with red lines), the ALV matches its perceptions with a topographical data base programmed into its computer. In a computer-generated image (right), light blue denotes a slope; lavender, mixed brush; red, an urban area; black, a road. Performing all the calculations necessary to replicate human sight is beyond the capabilities of any of today's computers.

The ALV's U. S. military sponsors, university contractors, and the Martin Marietta and Hughes Aircraft companies, hope to be able to build autonomous trucks, tanks, and mine detectors.

an inch, predicting the shortest path with superhuman accuracy. Named Ole, the robot practiced by finding a BB in a watermelon. Kwoh hopes that Ole may help, even perform, a variety of operations one day.

Other inventors are creating robotic caretakers and helpers. At the Veterans Administration Medical Center in Palo Alto, California, a robot arm hangs inert, its speech synthesizer emitting the unmistakable sounds of snoring. "Ready," says Stanford researcher Machiel Van der Loos. The snoring stops. "Attention; drink," he continues. "Drink," repeats a male computer-synthesized voice. The robot's prosthetic hand opens the desktop refrigerator, removes a glass of water with a straw, and aims it toward Mike's mouth. Designed to restore a measure of independence to quadriplegics, the aid can also heat soup in a microwave, feed and shave a patient, and brush his teeth. By the year 2000, predicts Larry Leifer, Stanford head of the project, robots working at such tasks may be as common as cars.

Robotics pioneer Joseph Engelberger would agree: "We're seeing the birth of a big, new industry" of service robots, he says. Engelberger founded one of the country's largest industrial robot manufacturers, Unimation, of Danbury, Connecticut. Recently he built HelpMate, a gofer for hospitals that works for subhuman compensation. "A nurse can send it to deliver a meal tray, take a urine sample to the lab." Engelberger's dream machine would be a $50,000 high-tech housekeeper that he hopes to start selling within the decade. This "ultimate robot" would cook, clean house, cut the grass, shovel snow, even fix washing machines and refrigerators, answering commands with a vocabulary of a few hundred words. Because of the "floppy object problem"—robots cannot handle fabric well—this maid will probably not make beds.

William "Red" Whittaker designs machines to work in less comfortable environments. "I envision complete automation of work like deep mining, underwater salvage, managing hazardous wastes," says Whittaker, of Carnegie Mellon University in Pittsburgh. He has presided over "breathing first life, first power" into two powerful robots that, guided by remote control, are cleaning up parts of the ruined Three Mile Island nuclear reactor building too radioactive for humans to enter. With colleagues, he has created prototypes for autonomous mining robots and is working on a NASA grant to help develop a six-legged robot to rove and study Mars.

Whittaker's machines still favor brawn over brain. But Hans Moravec, also at Carnegie Mellon, projects that the artificial mind will surpass ours in 20 to 40 years. "At that point," he says, "machines could do just about anything that a human being can do, including run our society." Moravec adds that people might be able to program, or "download," the contents of their mind into a computer with a robotic body, becoming essentially immortal. Moravec's views are controversial; for one thing, Marvin Minsky, a grand old man of artificial intelligence at MIT, would grant the human race another one to three centuries before androids catch up with us. But, warns another MIT pioneer of artificial intelligence, Joseph Weizenbaum, "There are components to human wisdom . . . that are not . . . machinable." If "a line dividing human and

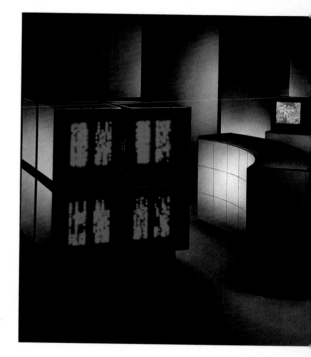

Experts in artificial intelligence— machines that "think" like humans—expect that computers like the Connection Machine (above) will not only drive autonomous vehicles but eventually transform their field. While most computers perform one calculation at a time, the Connection Machine works more like the human brain, putting 65,536 microcomputers to work simultaneously, and thus computing even faster than today's supercomputers.

The so-called parallel processors of tomorrow should be a thousandfold more powerful. Like the brain in computing capability, they may someday guide machines that see, think, move, and talk. Says Connection Machine inventor Daniel Hillis: "I would like to make a machine that can be proud of me."

machine intelligence" is not drawn, he fears, the dignity of man may be reduced to that of a clockwork automaton.

Today, though, biologists are exploiting computers to better understand human life. One person eager to use the fastest electronic helpers available is Leroy Hood, chairman of the Division of Biology at the California Institute of Technology. "What really drives the advance of science is not individual scientists per se," says Hood. "It's the development of new technologies, instruments, new kinds of chemistry." True to this vision, Hood has become a leader in devising machines to analyze the human animal, molecule by molecule.

At 50, Hood still runs five miles a day and goes rock- and mountain-climbing. He talks about science with the enthusiasm of, as he puts it, an adult Boy Scout. His grandest dream: using computer power and automated equipment to decipher the three billion bits of code in the total human genetic endowment, the human genome.

The genome project, which Hood calls "biology's equivalent of landing on the moon," is likely to cost some three billion dollars, take more than a decade to accomplish, and yield information that will require centuries to fully digest. From it scientists hope to learn no less than what makes all humans similar yet unique, how one cell creates a baby, how we age, what misspellings in the code cause the 3,000 genetic diseases, how we might correct genetic defects.

The answers lie locked inside deoxyribonucleic acid (DNA). DNA molecules spiral within every human cell and mastermind all biological activity by determining the proteins—biological building blocks—that each cell makes. DNA is made up of only four chemicals: molecules called bases and usually referred to as A, C, G, and T. Yet these four bases, arranged in a myriad of combinations, spell out enough messages to determine the shape, color, and biological function of virtually all forms of life on earth. Thousands of coupled bases make up one gene, the unit that codes for a protein. Yet an individual's genes account for only about 5 percent of the three billion bases in human DNA; the role of the rest remains mysterious.

Scientists have sleuthed out the location of some 2,000 human

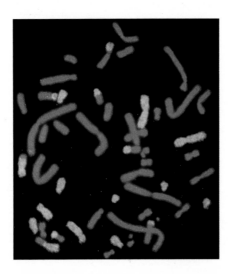

Geneticist Leroy Hood demonstrates the prototype of a gene analyzer, or sequencer, he developed at the California Institute of Technology. Laser beams help decipher the sequence of chemicals in the genetic code.

Human genetic matter, DNA, is bundled into 23 pairs of tiny rods called chromosomes. To pull out one chromosome, scientists sometimes fuse a human cell with a hamster cell; the hybrid will eject most of the human chromosomes, making the others relatively easy to spot (right). Dyed hamster DNA shows up orange; human, yellow.

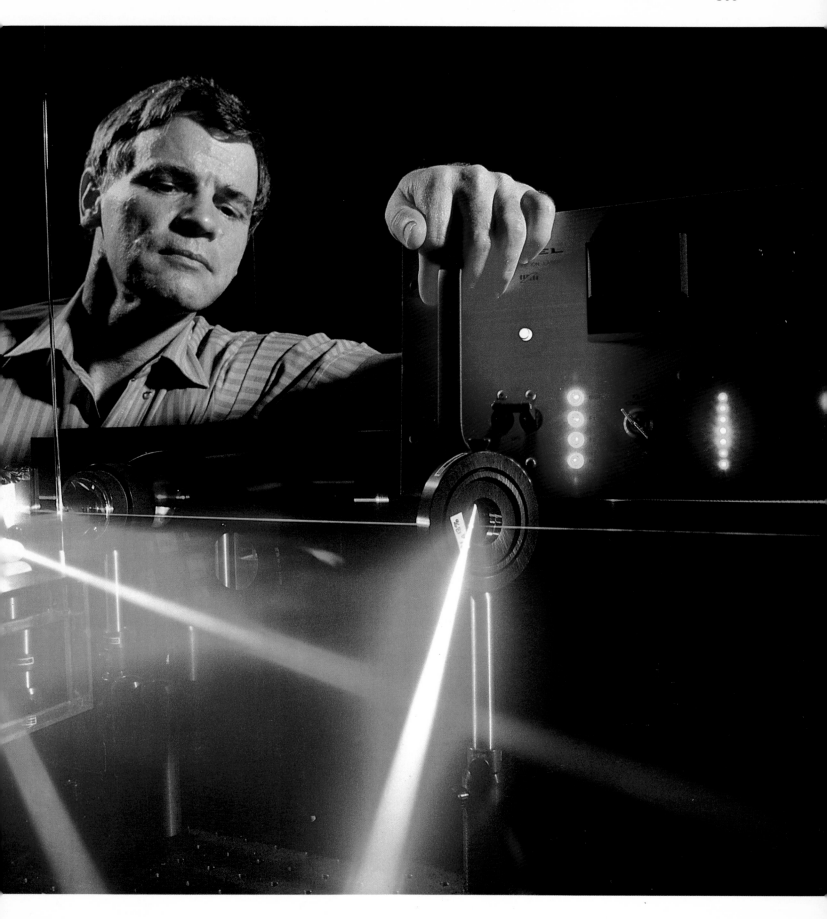

genes, gaining insights into—and diagnostic tools for—ailments including cystic fibrosis and forms of Alzheimer's disease, manic depression, and heart disease. Researchers plan to complete a physical map of human DNA so they can figure out what gene lies where in the rest of the immense genetic landscape. To do so, they will dice the genome into 250,000 overlapping pieces of DNA, then reassemble them, like solving a microscopic jigsaw puzzle.

Next, geneticists plan to analyze the bases' sequence. Sequencing by hand is a slow, tedious, and error-prone task that quickly burns out those who do it. Deciphering the entire code manually could take 7,000 years. Hood and his colleagues have created a semiautomated gene sequencer that mechanizes some of the many steps in preparing and analyzing DNA sequences. With a part-time technician, it can sequence in three days what it took five people two years to do by hand. Still, says Hood, the device is "just a Model T. In five years we hope to have a super Cadillac." Hood faces competition from others in the United States and abroad. Japanese industry, academia, and the government are coordinating an effort to sequence a million bases a day.

"Suppose I could tell you when you're born that you have a propensity to get diabetes," says Hood, "but if you did three things, you could decrease that propensity a hundredfold." Soon, Hood believes, scientists will develop a DNA analyzer that, from a finger-prick, could analyze DNA from blood cells and "give you a printout for all you're predisposed to," or at least problems like heart disease and arthritis that you might be able to control by diet, exercise, and other measures.

Furthermore, knowing the precise sequence of, say, the gene for Huntington's disease will allow scientists to determine just how it causes illness. If it is coding for a faulty protein, or none at all, researchers could use instruments that Hood and others have developed to reproduce the product of the normal gene and administer it as a drug.

But is it possible to know too much? Critics fear that genetic prophecies might prompt discrimination by health and life insurance companies as well as prospective employers. Genetic screening, already performed for a few conditions, could also provoke heartbreaking decisions for couples whose unborn child may have a disability. "The technology itself is value free," insists Hood. "But we need to create forums which draw from all elements of society—scientists, ethicists, ministers, businessmen—to make recommendations" on how to handle new genetic knowledge. In the meantime, the passion to discover motivates those spurring the genome project forward.

On the brink of the 21st century, the alliance of discoverer and inventor is closer than ever before. A new finding can inspire an innovation within months, and inventions can in turn lead with galloping speed to the next discovery. Yet no matter how advanced science becomes, discoverers and inventors still break through technological barriers by doggedly tracking an idea—be it logical or unlikely—or by examining the serendipitous. And the greatest challenge may still be, in the words of a researcher of superconductivity, "How can we do something that no one has yet imagined?"

Chemical bases dubbed A, C, G, and T form the active ingredients of DNA. Thousands of these bases make up each of an individual's 100,000 genes. Their order spells out the genetic blueprint of every life. A technician in Leroy Hood's lab (right) pours samples of DNA into the gene sequencer. Under laser light, fluorescent dyes (far right) will flag individual bases with characteristic colors, enabling the machine to decode the bases' order. With this information, scientists hope to find causes and cures for genetic ills. Many believe we will learn as much about human biology in the next twenty years as we have in the past two millennia.

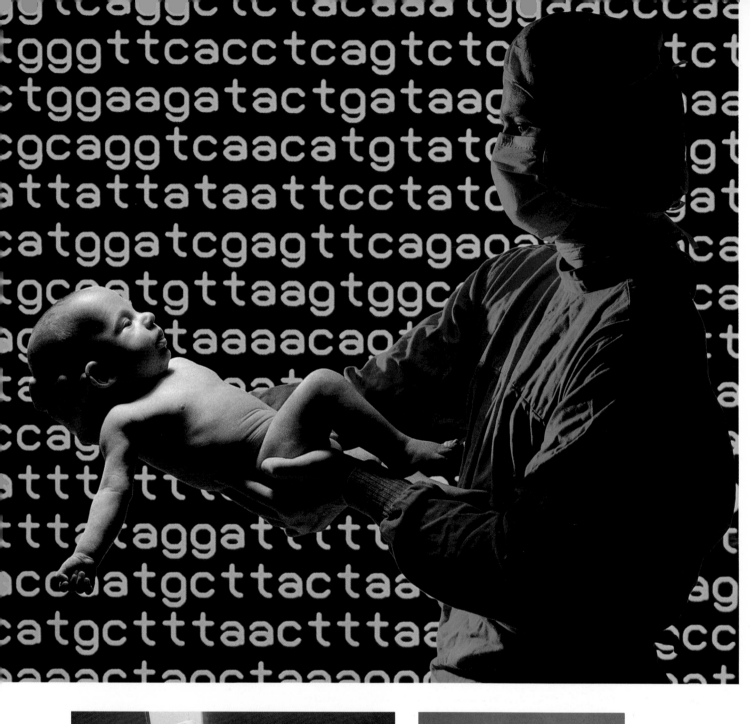
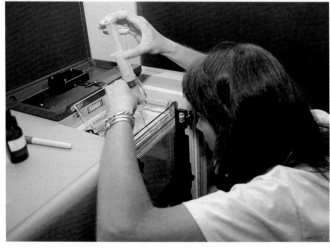

About the Authors

Thomas B. Allen is a writer of both fiction and nonfiction. His recently published books include *Merchants of Treason* (with Norman Polmar) and *War Games*. A former National Geographic staff member, he was the editor of *We Americans* and *Images of the World: Photography at the National Geographic*.

Daniel J. Boorstin served as Librarian of Congress for 12 years, until 1987. He had been director of the Smithsonian's National Museum of American History and the institution's Senior Historian, following 25 years at the University of Chicago. He is the author of 18 books on American and world history, including *Hidden History*, *The Image*, the best-selling *The Discoverers*, and a trilogy, *The Americans*—volumes of which won Bancroft, Francis Parkman, and Pulitzer Prizes.

Tom D. Crouch, former curator of aeronautics at the Smithsonian's National Air and Space Museum, is the author of a number of books and articles on aerospace history, including *A Dream of Wings: Americans and the Airplane, 1875–1905; The Eagle Aloft: Two Centuries of the Balloon in America;* and, most recently, *Wilbur and Orville Wright: A Biography.* He is now chairman of the Department of Social and Cultural History at the Smithsonian's National Museum of American History.

H. Garrett DeYoung writes about international issues in science and technology for a number of publications, including *Bio/Technology, High Technology Business,* and *Inc.* Formerly a senior editor specializing in chemistry and life sciences at *High Technology,* he is the author of *The Cell Builders,* an account of the development of monoclonal antibodies and their use in cancer treatments.

Carole Douglis specializes in writing about science and technology for national publications, including the *Atlantic, Omni,* and *Psychology Today.* She has also contributed to other National Geographic books and has acted as a consultant for the National Academy of Sciences.

Robert Friedel is a writer and professor in the history of technology at the University of Maryland, College Park. His books include *Pioneer Plastic: The Making and Selling of Celluloid* and *Edison's Electric Light: Biography of an Invention.* A contributing editor of *American Heritage of Invention and Tech-*

nology, he recently acted as consultant for the Smithsonian exhibition "A Material World," for which he wrote the catalog.

Stephen S. Hall has written science and travel articles for many national magazines, including *Smithsonian,* National Geographic *Traveler,* the *New York Times Magazine, Omni,* and *Hippocrates.* His recently published book, *Invisible Frontiers,* is a popular history of the origins of genetic engineering.

T. R. Reid is a reporter on the National News staff of the *Washington Post.* He also writes a nationally syndicated column, "The Computer Report." He became interested in the microchip after building his own personal computer and later authored *The Chip: How Two Americans Invented the Microchip and Launched a Revolution,* published in 1984.

Richard Rhodes is the author of eight books and more than sixty magazine articles. He was awarded a 1987 National Book Critics Circle Award, a National Book Award and, in 1988, a Pulitzer Prize—the so-called triple crown—for *The Making of the Atomic Bomb.* He is currently a fellow at the Center for International Studies at the Massachusetts Institute of Technology.

Margaret Sedeen is a National Geographic book editor, whose most recent volumes were *Mountain Worlds* and *Great Rivers of the World.* Also a writer, she has covered topics ranging through biography, science, social history, and geography. She is especially interested in 19th-century history of science and social history.

Fred Strebeigh teaches writing at Yale University and has written articles about social history for national publications, including *Smithsonian, American Heritage,* the *New Republic,* and *Audubon.* He has journeyed to a number of sites significant in the history of radio including Signal Hill in Newfoundland, where the first transatlantic wireless message was received.

Acknowledgments

We wish to thank the many individuals, groups, and institutions who helped in the preparation of *Inventors and Discoverers: Changing Our World.*

Our special thanks go to David K. Allison, Uta Merzbach, Eugene Ostroff, and Elliot N. Sivowitch of the Smithsonian Institution's National Museum of American History; Herbert L. Anderson; Silvio A. Bedini, Smithsonian Institution Libraries; Robert V. Bruce; James J. Flink of the University of California, Irvine; Donald S. Lopez of the Smithsonian Institution's National Air and Space Museum; Emilio Segrè; and Paul Spehr of the Library of Congress.

We are also indebted to Edward D. Aebischer, Oak Ridge National Laboratory; John-Michael Albert, University of Houston; Robert Alter and Rebecca Kenney, Polaroid Corp.; Frederic N. Barry and George K. Durfey, Foothill College; Gwen Bell and Linda Holekamp, The Computer Museum, Boston; Thomas O. Binford and Bruce G. Buchanan, Stanford University; Mr. and Mrs. Bradford Blackman; Betsy Brayer; Julie Carroll, Peter O'Connell, and Don B. Olsen, University of Utah; Thomas Chang, McGill University; Lillian Christmas, The Roland Institute for Science; Kevin Compton, Robert Ford, and Ruth Lombardi, AT&T Bell Laboratories; Philip Condax, Rachel Stuhlman, Morgan Wesson, and David Wooters, International Museum of Photography at George Eastman House; Timothy J. Countryman, Young Morse Historic Site; David R. Crippen, Henry Ford Museum and Greenfield Village; Susan Davis, Stanley Museum; Larry Deaven, Don Feldman, Gary Glatzmaier, Walter Goad, Roger Meade, Bill Jack Rodgers, and Mollie Rodriguez, Los Alamos National Laboratory; Robert Dreesen, Richard D. Horigan, Jr., Peter Jakab, Robert Mikesh, Larry Wilson, and Howard Wolko, National Air and Space Museum; Chris Driscoll, Children's Hospital National Medical Center, Washington, D. C.; John Duffy; and Sheila Edelheit, Medical College of Hampton Roads.

We are grateful to Jon B. Eklund, Bernard S. Finn, Paul Forman, Stanley Goldberg, Robert S. Harding, Ann M. Seeger, and Roger B. White, National Museum of American History; K. G. Engelhardt, Mark Friedman, Raj Reddy, and Chuck Thorpe, Carnegie Mellon University; Shari Folz, Pacific Data Images; Liliana Gagliardi; Alan Gevins, EEG Systems Laboratory; David Gibson, Eastman Kodak Co.; Ron Grantz, National Automotive History Collection; Laurie Hansen, Con

Illustrations Credits

Edison Co.; William Herrman and David Y. Tseng, Hughes Aircraft Co.; John Hopfield, Sarah Kelley, Barbara Otto, and Jane Sanders, California Institute of Technology; Stephen H. Howell, University of California, San Diego; Charles Hufnagel, Georgetown University; Jay Hurley, Rumsian Society; International Tesla Society, Inc.; David Israel and Stanley J. Rosenschein, SRI International; Robert Joy, Department of Defense; Rainer Karnowske and Michael Rau, Mercedes-Benz Museum, Stuttgart, W. Germany; Evelyn Karson, Janice Lazarus, Daniel Masys, and John Parascandola, National Institutes of Health; Gus Kayafus, Palm Press; Arthur E. Koski, Evan McCollum, and David Weltch, Martin Marietta Denver Aerospace; Al Krisciunas, Argonne National Laboratory; Albert Kuhfeld, Bakken Library; Edward Lofgren, Doug McWilliams, and Dorothy Tuley, Lawrence Berkeley Laboratories; Eugene Lowenthal, Microelectronics and Computer Technology Corp.; and Maurice J. Mahoney, Yale University School of Medicine.

Our thanks also go to Mary Manley and Richard Willett, Goodyear Tire and Rubber Co.; Katherine Martyn, University of Toronto; Barbara McCandless, University of Texas at Austin; Sally Merryman, Texas Instruments Inc.; Jack Minker, University of Maryland; Edward Mitchell, Jr., INVESCO Capital Management; Monique Montague; James Morgan and John Swenson, Cray Research, Inc.; Ruth Murray, University of Akron; Yukihiko Nosé, Cleveland Clinic Foundation; Stephen H. Ober, Motion Control, Inc.; George O'Hare, New England Bell Telephone Co.; Eric Olson, Edison National Historic Site; William Pierce, Pennsylvania State University; Frank Ruddle, Yale University; Scientists' Institute for Public Information; George Seidell, Colorado State University; Ruth Shoemaker, General Electric Co.; Arnold Shuman, Mercedes-Benz of North America; Phyllis Smith, David Sarnoff Research Center; Rudi Stern, Let There Be Neon, Inc.; Tesla Memorial Society Inc.; David von Endt, Smithsonian's Museum Support Center; Spencer Weart, American Institute of Physics; Daniel Weiss; and Patrick H. Winston, Massachusetts Institute of Technology.

In addition, we gratefully acknowledge the help of Luis Marden and Volkmar Wentzel; Valerie Mattingley and Jennifer Moseley, National Geographic United Kingdom Office; John Morris and Candace Roper, National Geographic Paris Office; and the National Geographic Administrative Services, Illustrations Library, Library, Photographic Laboratory, Records Library, and Travel Office.

Abbreviations for terms appearing below: (t)-top; (c)-center; (b)-bottom; (l)-left; (r)-right; ENHS-Edison National Historic Site; LC-Library of Congress; NASM-National Air and Space Museum; NGP-National Geographic Photographer; NGS-National Geographic Staff; NMAH-National Museum of American History.

Cover stamping, Carl R. Mukri. Pages 1 and 3, Michel Tcherevkoff. Deluxe edition, page 3, hologram, American Bank Note Holographics, Inc.

Discoverers & Inventors
6, Michel Tcherevkoff. 8-9, Michael Freeman. 10-11, Musée Condé, Chantilly, France: Giraudon/Art Resource. 12(l), Fred J. Maroon. 12(c), Sisse Brimberg. 12(r), Victor R. Boswell, Jr., NGP. 12-13, MacDonald/Orbis: Angelo Hornak. 13(l), Bates Littlehales, NGP. 13(c), Joseph D. Lavenburg, NGP. 13(r), Ira Block.

The Power of Steam
14, Michel Tcherevkoff. 16-17, James A. Sugar. 18, "Watt discovering condensation of Steam," by Marcus Stone; Galerie George, London: The Bridgeman Art Library. 19(t), National Museums and Galleries on Merseyside: Walker Art Gallery. 19(b) & 20, North Wind Picture Archives. 21, Daniel Meadows. 22, The Science Museum, London. 23, Harper's Weekly, 1876: LC. 24, Centennial Mirror published by the American Oeleograph Co., 1876: LC. 24-25, Grant Heilman Photography. 26-27, "Richard Trevithick's Railroad Euston Square 1809," by Thomas Rowlandson; The Science Museum, London: The Bridgeman Art Library. 27, "Richard Trevithick," by John Linnell; The Science Museum, London: The Bridgeman Art Library. 28, "George Stephenson and Family," by William Lucas, 1861: The Science Museum, London. 29, "Locomotion," by Terence Cuneo, 1949: Weidenfeld and Nicolson Archives. 30, Lithograph by J. C. Bourne, 1838; National Railway Museum: Tinted by Michael A. Hampshire. 31, Lithograph by S. Russell, 1849; The Science Museum, London: Tinted by Michael A. Hampshire. 32-33, C. P. Lewis. 34-35, Arthur Lidov. 36, Miniature by Fulton after Benjamin West; The New-York Historical Society: Victor R. Boswell, Jr., NGP. 37(t), The New-York Historical Society: Victor R. Boswell, Jr., NGP. 37(b), The New York Public Library. 38-39, LC. 39, Collection of Leonard V. Huber. 40, Lithograph by Kummel and Forster: The Bett-

mann Archive. 41, National Maritime Museum. 42, BBC Hulton Picture Library. 43, Charles E. Rotkin, Photographs for Industry.

The Age of Electricity
44, Michel Tcherevkoff. 46-47, Con Edison Co. 48-49, The Bakken, Minneapolis. 49, by L. Guiguet from Les Merveilles de La Science, L. Figuier, Paris, 1867-1870, vol. 1: British Library. 50(t), The Royal Institution, London. 50(b), Smithsonian Institution: Victor R. Boswell, Jr., NGP. 50-51, Watercolor by Harriet Moore, 1852: The Royal Institution, London. 52, Illustrated London News Picture Library: Tinted by Michael A. Hampshire. 53, Con Edison Co. 54, National Academy of Design. 54-55, Brady Collection, LC. 55, Young Morse Historic Site, Locust Grove: Breton Littlehales. 56, U. S. Department of the Interior, National Park Service, ENHS. 57-59, ENHS: Breton Littlehales. 59(b), the New York Times. 60-61, ENHS. 60(bl) ENHS: Breton Littlehales. 62, General Electric Co. 63(t), Museum of the City of New York. 63(b), Con Edison Co. 64, Kenneth M. Swezey Collection, Archives Center, NMAH. 65, Burndy Library, Norwalk, CT. 66-67, W. Bertsch, H. Armstrong Roberts. 67, Niagara Mohawk Power Corp. 68, American Telephone & Telegraph Co. 69, New England Bell Telephone Co.: Breton Littlehales. 70-73, Bell Family Collection, LC. 74(l), L. W. Harris. 74(r), Keystone. 75, American Telephone & Telegraph Co. 76, Brown Brothers. 77, The Bettmann Archive. 78, Roger-Viollet, Paris. 78-79, Grafton M. Smith, Image Bank.

On Wheels & Wings
80, Michel Tcherevkoff. 82-83, Underwood and Underwood. 83, The Bettmann Archive. 84, Mercedes-Benz of North America. 85(l), Adam Woolfitt. 85(r), Daimler-Benz AG. 86 (both), Mercedes-Benz of North America. 87, Adam Woolfitt. 88-89, Brown Brothers. 90(t), Automotive Collection, Detroit Public Library. 90(b), The Peter Roberts Collection. 91, Martin Sandler Productions. 92-97, Henry Ford Museum and Greenfield Village. 98(t), Con Edison Co. 98(b), NMAH. 99, The Bettmann Archive/BBC Hulton. 100, The Granger Collection. 101, R. W. Wood. 102-103, LC. 104, NASM. 104-105, LC. 106(t), Christopher Springmann. 106(b), NASM. 107, West Indian Aerial Express. 108-109, Paul E. Garber Facility, Smithsonian: Breton Littlehales. 110(b), FPG. 110-114, NASM. 114-115, the Rolls-Royce Magazine. 116, NASM. 116-117, Dean Conger for NASA.

A World of New Materials
118, Michel Tcherevkoff. 120-121, Michael Skott. 122-123, Stanley Meltzoff. 124, The Goodyear Tire and Rubber Co., Akron: Breton Littlehales. 125, Willard R. Culver. 126(l), BASF Corp. 126(r), by Sir Arthur S. Cope, 1906: National Portrait Gallery,

London. 127, Analytical Laboratory, Smithsonian: Breton Littlehales. 128(t), Archives Center, NMAH. 128 & 129, Brian Hagiwara. 130(t), Archives Center, NMAH. 130(b) NMAH: Bill Ballenberg. 131(t), Archives Center, NMAH. 131(b), NMAH: Bill Ballenberg. 132-133 (both), The Hagley Museum and Library. 134, Willard R. Culver. 135, Ted Polumbaum for *Life,* Time, Inc. 136, Ralph Crane for *Life,* Time, Inc. 136-137, Tupperware Home Parties. 138-139, Dan Nerney/DOT. 140(t), Aluminum Company of America. 140(b), James L. Amos, NGP. 141, Musée du Compiègne, France: James L. Amos, NGP. 142, James L. Amos, NGP. 143, David S. Boyer, NGS. 144, Willard R. Culver. 144-145, Elliott Kaufman. 146, Department of Archives and Records Management, Corning Glass Works. 146-147, Breton Littlehales.

Capturing the Image
148, Michel Tcherevkoff. 150-151, General Electric Co. 152, Gernsheim Collection, Harry Ransom Humanities Research Center, University of Texas at Austin. 153(l), International Museum of Photography at George Eastman House. 153(tr & b), International Museum of Photography at George Eastman House: Breton Littlehales. 154-155, International Museum of Photography at George Eastman House. 155, The Science Museum, London. 156(tl), Gernsheim Collection, Harry Ransom Humanities Research Center, University of Texas at Austin. 156-157, International Museum of Photography at George Eastman House: Breton Littlehales. 158, The Science Museum, London. 158-159, Gernsheim Collection, Harry Ransom Humanities Research Center, University of Texas at Austin. 160, Eastman Kodak Co. 161, International Museum of Photography at George Eastman House: Breton Littlehales. 162(b), International Museum of Photography at George Eastman House: Breton Littlehales. 162-163, International Museum of Photography at George Eastman House. 164(t), Collection of Julien Gerardin-Christian Debize, Nancy, France. 164(b), Charles Martin. 165(t), International Museum of Photography at George Eastman House. 165(c), W. Robert Moore. 165(b), Eastman Kodak Co. 166-167, Harold E. Edgerton. 168(both), Polaroid Corporate Archives. 169, Marie Cosindas. 170(t), Gernsheim Collection, Harry Ransom Humanities Research Center, University of Texas at Austin. 170(b), International Museum of Photography at George Eastman House: Breton Littlehales. 170-171, Dr. George E. Nitzsche Collection, Philadelphia Museum of Art: Graydon Wood. 171(t), The Illustrated London News Picture Library. 172(t), Museum of Modern Art Film Stills Library. 172(b), International Museum of Photography at George Eastman House: Breton Littlehales. 173, Edison Collection, LC. 174(t), The Bettmann Archive. 174-175,

International Museum of Photography at George Eastman House: Breton Littlehales. 176, The Granger Collection. 176-177, MGM/Kobal Collection. 178-179, Museum of Modern Art Film Stills Archive. 180, MGM/Kobal Collection. 181, International Museum of Photography at George Eastman House Still Collection.

Messages by Wireless
182, Michel Tcherevkoff. 184-185, Danny Lehman. 186, RCA. 187-188, GEC-Marconi. 189, Amministrazione Postale Italiana, Museo Storico, Rome: Adam Woolfitt. 190, Smithsonian Institution: IEEE Center for the History of Electrical Engineering. 190-191, GEC-Marconi. 192(t), Foothill Electronics Museum. 192(b), NMAH: Bill Ballenberg. 193, GEC-Marconi. 194(t), Press-Cliché Russian News Photo Agency. 194(b), NMAH: Bill Ballenberg. 195, Archives Center, NMAH. 196, Perry Cragg. 197, Culver Pictures. 198-199, Robert Hunt Library. 199(both), Imperial War Museum. 200(both), RCA. 201(l), IEEE Center for the History of Electrical Engineering. 201(r), RCA. 202-203, J. Baylor Roberts. 204, National Archives. 205, UPI/Bettmann Newsphotos. 206, Yale Joel for *Life,* Time, Inc. 207, RCA. 208-209, Willard R. Culver. 210-211, Fred Ward, Black Star. 211, Robert B. Goodman. 213, Vincent Di Fate.

Power Particles
214, Michel Tcherevkoff. 216-217, Los Alamos National Laboratory. 218(t), Science Photo Library. 218(b), Deutsches Museum, Munich, W. Germany. 219, Brown Brothers. 220-221, Lawrence Berkeley Laboratory. 221(l), Lawrence Berkeley Laboratory: Victor R. Boswell, Jr., NGP. 222-223, Thomas J. Abercrombie, NGS. 224, Institut Curie: AIP Niels Bohr Library. 225(l), Lawrence Berkeley Laboratory: Science Photo Library. 225(r), AIP Niels Bohr Library. 226-227, Hugo Jaeger for *Life,* Time, Inc. 227, UPI/Bettmann Newsphotos. 228, Herbert L. Anderson. 229(t), "Birth of the Atomic Age," by Gary Sheahan, 1957: Chicago Historical Society. 229(b), Emilio Segrè. 230-231, Los Alamos National Laboratory. 232, Fred J. Maroon. 233, Charles Harbutt, Archive. 234-235, Yann Arthus-Bertrand, Peter Arnold, Inc. 236, Charles Harbutt, Archive. 237, Howard Sochurek. 238-239, Danny Lehman. 240(t), Edwin B. Bruening. 240(b), NMAH. 240-241, Charles O'Rear, West Light. 242, Fred Ward, Black Star. 243(l), Charles Feil, Stock Boston. 243(r), Charles O'Rear, West Light. 244(l), Philippe Plailly, Science Photo Library. 244(r), William James Warren, West Light. 245, Roger Ressmeyer, Starlight.

Computers & Chips
246, Michel Tcherevkoff. 248-249, Erich Hartmann, Magnum. 250, NMAH: Bill Ballenberg. 251, Associated Press/Wide World.

252(t), *Time* magazine. 252-253, Fred J. Maroon. 254-255, The Computer Museum. 256, Wayne Miller, Magnum. 257, Texas Instruments. 258, Charles O'Rear, West Light. 258-259, Tom Tracy, Black Star. 259(t), National Semiconductor. 259(b), Erich Hartmann, Magnum. 260-261, Kay Chernush. 262, Erich Hartmann, Magnum. 263, Hank Morgan, Rainbow. 264-265, Joseph Rodriguez, Black Star. 265, E. Miyazawa, Black Star. 266, Douglas Greer, EEG Systems Laboratory, San Francisco. 266-267, William Strode. 267(both), Douglas Greer, EEG Systems Laboratory, San Francisco. 268-269, Erich Hartmann, Magnum. 269, Dan McCoy, Rainbow. 270-271, Brian R. Wolff. 272, Pacific Data Images: Rob Daly, VideoStill Imaging, Los Angeles. 273, Pacific Data Images.

Engineering Life
274, Michel Tcherevkoff. 276-277, Michael Lang, Visum. 278(l), "Pasteur dans son laboratoire," by Albert Edelfelt, 1884: Musée Pasteur, Institut Pasteur. 278(r), Roger-Viollet, Paris. 279(t), Brown Brothers. 279(b), Georg Gerster. 280, National Archives. 281(l), Black Star. 281(r), Brown Brothers. 282, March of Dimes Birth Defects Foundation. 283, Bill Eppridge for *Life,* Time, Inc. 284-285, March of Dimes Birth Defects Foundation. 285, George Kew for *Life,* Time, Inc. 286, Ralph Morse for *Life,* Time, Inc. 287(t), Associated Press. 287(b), Science Photo Library. 288(t), Erich Hartmann, Magnum. 288(b), Willem J. Kolff. 289, Dan McCoy, Rainbow. 290-291, Erich Hartmann, Magnum. 290(b), Motion Control, Inc. 292(l), R. Langridge and Dan McCoy, Rainbow. 292(r), Cold Spring Harbor Laboratory. 293, Ted Spiegel, Black Star. 295(l), Steve Northup, Black Star. 295(tr), Ted Spiegel, Black Star. 295(br), Robert Hammer and Ralph Brinster, University of Pennsylvania, School of Veterinary Medicine. 296, Keith V. Wood. 297(both), Colorado State University. 298, Hank Morgan, Rainbow. 298-299, Lennart Nilsson. 299, Hank Morgan, Rainbow.

Technology of Tomorrow
300, Michel Tcherevkoff. 302, IBM. 303, Robert W. Madden, NGS. 304-306(t), Erich Hartmann, Magnum. 306(b), Hughes Aircraft Co. 307, Thinking Machines Corp: Stephen Grohe. 308, Los Alamos National Laboratory: Evelyn Campbell. 308-309, Roger Ressmeyer, Starlight. 310-311, Michael Lange, Visum. 311(bl & br), Erich Hartmann, Magnum.

Index

Type composition by the Typographic section of National Geographic Production Services, Pre-Press Division. Color separations by Chanticleer Co., Inc., New York, N.Y.; The Lanman Companies, Washington, D. C.; Lincoln Graphics Inc., Cherry Hill, N.J. Printed and bound by R. R. Donnelley and Sons Co., Chicago, Ill. Paper by Mead Paper Co., New York, N.Y.

Library of Congress CIP Data

Inventors and discoverers. changing our world/prepared by National Geographic Book Service. — 1st ed.
    p.   cm.
    Includes index.
    ISBN 0-87044-751-3 (alk. paper): $19.95. ISBN 0-87044-752-1 (delux: alk. paper): $29.95. ISBN 0-87044-753-X (lib. bdg.: alk. paper): $21.95
    1. Inventions—History.   I. National Geographic Book Service.
T18.I57 1988
609—dc19
                      88-22375
                      CIP